Intelligent Technical Systems

Lecture Notes in Electrical Engineering

Volume 38

For further volumes:
http://www.springer.com/series/7818

Natividad Martínez Madrid · Ralf E.D. Seepold
Editors

Intelligent Technical Systems

Springer

Editors
Prof. Dr. Natividad Martínez Madrid
Universidad Carlos III Madrid
Depto. Ingenieria
Avenida Universidad, 30
28911 Leganes
Spain
nati@it.uc3m.es

Prof. Dr. Ralf E.D. Seepold
Universidad Carlos III Madrid
Depto. Ingenieria
Avenida Universidad, 30
28911 Leganes
Spain
ralf@it.uc3m.es

ISBN 978-1-4020-9822-2 e-ISBN 978-1-4020-9823-9

DOI 10.1007/978-1-4020-9823-9

Library of Congress Control Number: 2009920106

Printed on acid-free paper

9 8 7 6 5 4 3 2 1

springer.com

Preface

Intelligent Technical Systems are electronic devices in which one or more networked components are located. Nowadays, the sectors of automotive, medical/e-Health or multimedia show interesting developments. Other well-known sectors like home or building automation also introduce new concepts in this area.

Intelligent Technical Systems are characterized by a strong interaction with their environment. Several of these systems require mobility support. For example, systems with ubiquitous computing capabilities require a complex design of different interfaces. The integration of human-machine interfaces needs to be considered in special purpose systems, like devices for dependent people with specific needs.

The book *Intelligent Technical Systems* provides an overview on several related fields of applied research, like multimedia systems, embedded programming, middleware platforms, sensor networks/autonomous systems and applications for intelligent engineering. Each area is covered by a separate part of the book.

This book supports application engineers and researchers to get introduced into the topic of Intelligent Technical Systems with the help of concrete examples covering the design and implementation phase.

Madrid, Spain
December 2008

Natividad Martínez Madrid
Ralf E.D. Seepold

Contents

Contributors

Henk Aarts
Jon Altuna
Juan Antonio Álvarez
Björn Andersson
Jesús Barba
Maitane Barrenechea
Marco Caldari
Paulo Carvalho
Antón Civit
Fabrizio Concettoni
Massimo Conti
Javier Del Ser
Julio Dondo
Wilfried Elmenreich
Alejandro Fernández-Montes
Gilles Grimaud
Bernhard Huber
Mario Ibáñez
Willibald Krenn
Markus Kucera
Gabriel Jiménez
Jeong-Bae Lee
Alejandro Linares-Barranco
Juan Carlos López
Alberto Los Santos
Carlos Luján-Martínez
Kevin Marquet
Jaime Martín
Javier Martínez
Natividad Martínez Madrid
Mikel Mendicute
Lorenzo Morbidelli
Francisco Moya
Fortunato Nocera
Roman Obermaisser
Simone Orcioni
Juan Antonio Ortega
Nuno Pereira
Daniele Petraccini
Alvaro Reina
Fernando Rincón
Franco Ripa
Maximilian Rosenblattl
Imran Sabir
Jesús Sáez
Pilar Sanz
Ralf Seepold
Gerald Steinbauer
Eduardo Tovar
Athanassios Tsakpinis
Bernhard Turban
Michael Vetter
David Villa
Félix J. Villanueva
Andreas Wolf
Christian Wolff
Franz Wotawa
Sang-Young Cho

Reviewers

Massimo Conti
Università Politecnica delle Marche, Ancona, Italy

Wilfried Elmenreich
University of Klagenfurt, Austria

Natividad Martínez Madrid
Universidad Carlos III de Madrid, Spain

Ralf E.D. Seepold
Universidad Carlos III de Madrid, Spain

Part I

Multimedia Systems

Chapter 1

Smart Wireless Image Sensors for Video Surveillance

Massimo Conti and Simone Orcioni
DIBET, Università Politecnica delle Marche, Ancona, Italy, m.conti@univpm.it

Abstract This chapter presents an analysis of image processing performance analysis of wireless image sensors networks for video surveillance. The dependence of image quality, network throughput and channel noise sensitivity with image enhancement algorithms, image compression and wireless protocols have been investigated. The objective of the work is to give useful guidelines in the design of image wireless networks over low cost, low power, low rate Bluetooth and Zigbee wireless protocols.

Keywords Image processing, Video surveillance, Wireless network, Bluetooth, Zigbee.

1.1 Introduction

Nowadays image sensors based on CMOS technology are very common in portable devices. They present many useful characteristics offering low-power and low-cost features as well as high quality. The problem of improving the quality of images acquired in critical illumination conditions is interesting for consumer video, photo camera applications and for video surveillance applications. As a consequence, there is a growing interest on efficient and low complexity implementations of nonlinear algorithms for image enhancement [1].

Research on sensor networks was originally motivated by military applications; however the availability of low-cost sensors and communication networks has resulted in the development of many other potential applications, like Infrastructure Security, Environment and Habitat Monitoring, Health Monitoring, Industrial Sensing and Traffic Control [2,3].

N. Martínez Madrid, R.E.D. Seepold (eds.), *Intelligent Technical Systems*, Lecture Notes in Electrical Engineering 38, DOI 978-1-4020-9823-9_1,

This contribution focuses on the analysis of image processing algorithms for wireless networks of smart low power image sensors for video surveillance.

A sensor network design is influenced by many factors, which include fault tolerance, scalability, production costs, operating environment, hardware constraints and power consumption [4–6]. These factors are important because they serve as a guideline to design the protocol, the network topology, the processing performed by each sensor, the data to be transmitted, the application layer, the hardware/software implementation of the transmitter and receiver and of the image processing.

Video surveillance recently is becoming one of the promising applications for wireless networks [7–18]. Hengstler [16] presented a tool based on Matlab for investigation and development of applications for wireless sensor networks.

In Hengstler et al. [17], a smart camera architecture is presented for local processing, with the aim of facilitating distributed intelligent surveillance. The paper focuses mainly on the efficient data link layer utilizing IEEE 802.15.4 MAC for enabling data exchange between smart cameras.

The IBM smart surveillance system [18] is able to monitor, manage surveillance data, detect, track and classify objects, with the possibility to manage in real time the surveillance system on the Web. But the aspect of low power for implementation on battery supplied wireless cameras is not addressed in the system.

The wireless sensor node can usually be equipped with a limited power source. Sensor node lifetime, therefore, shows a strong dependence on battery lifetime. The switching off of few nodes in a sensor network can cause significant topological changes and might require a re-organization of the network. Hence, power management requires additional importance. For these reasons researchers are currently focusing on the design of power-aware protocols and algorithms for sensor networks.

The main task of a sensor node in a sensor field is to detect events, perform quick local data processing, and then transmit the data. Power consumption can hence be divided into three domains: sensing, communication, and data processing.

Sensing power varies with the nature of applications. Sporadic sensing might consume less power than constant event monitoring. The complexity of event detection also plays a crucial role in determining energy dissipation.

The main specifications to be considered in the design of image sensor wireless networks are the quality of the image to be transmitted, the number of images per second that can be transmitted, the power dissipated for image processing and transmission and the insensitivity with noise in the channel.

Image processing algorithms, image compression algorithms, sensor node architecture, wireless protocol from application layer to physical layer, network topology must be analyzed and all parameters should be tuned in a system level simulation of the complete sensor network in order to optimize the performances of the complete system.

Image compression in the sensor reduces data to be transmitted, but this causes a reduction in image quality and a greater sensitivity to noise. Conversely, image processing can be performed to improve quality before compression in the sensor, but the processing will increase the power dissipated by the sensor itself and the compression ratio will be reduced with a consequent increment in the transmission power and sensitivity to noise. Image processing on the receiver side, after applying compression, does not allow the same quality, but it reduces power dissipation on the sensor node.

This work presents an analysis of the dependence of image quality, network throughput, and channel noise sensitivity with image enhancement algorithms, JPEG image compression parameters and wireless protocols.
In particular ZigBee and Bluetooth wireless standard will be considered.

In Section 1.2 the architecture of the smart wireless sensor node is presented, and some results obtained with Matlab simulations are shown. Section 1.3 briefly presents the ZigBee and Bluetooth wireless standards and shows an analysis of the performance of a wireless sensor network obtained with Matlab and SystemC.

1.2 Smart Wireless Image Sensor

The processing flow of a Smart Wireless Image Sensor is reported in Fig. 1.1. Before JPEG compression, the image is processed by an image enhancement algorithm, then the data are coded following the wireless protocol, and noise in the channel can be added. Image enhancement is applied to improve the quality of the image for successive image recognition on the central control station.

In video surveillance a key aspect is the improvement in image quality in bad illumination conditions. For example, it is extremely important to identify details in a not illuminated part of the image, when another part of the image is extremely illuminated. A lot of work is devoted to this kind of problem which takes inspiration from the retinex algorithm of Land [19]. Following the retinex approach the image $I(x,y)$ is considered as the product of two terms: the illumination $L(x, y)$ and the reflectance $R(x, y)$. The retinex algorithm requires the calculus of very complex functions, such as logarithm, exponential function and division, functions that are very expensive in terms of power dissipation and hardware complexity.

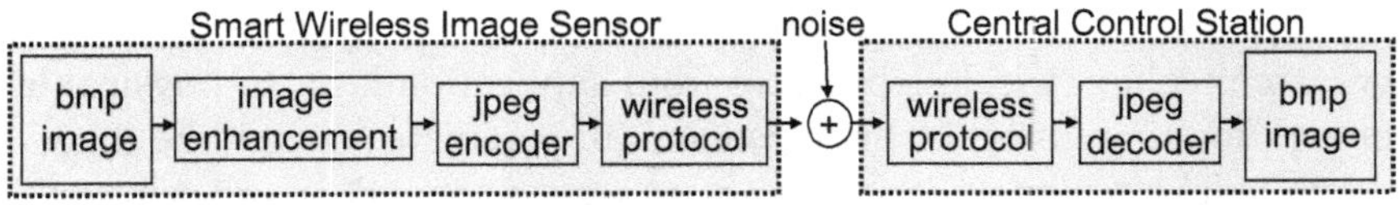

Fig. 1.1 Signal processing flow in the smart wireless image sensor

Recently a simplified version of the retinex algorithm has been developed [20] with a strong reduction in the complexity, but acceptable quality in image processing. The low complexity and low power dissipation of this simplified algorithm, that we will call "nonlinear stretching" in the rest of the paper, allow the implementation of image processing in the sensor.

The quality of the image can be improved with nonlinear stretching algorithm, even in case of image compression, if image processing is performed before compression.

Widely used image compression algorithms are JPEG and JPEG2000. The effect on the image quality with JPEG and JPEG2000 compression over a ZigBee network has been investigated [21], with the conclusion that JPEG2000 reaches better quality, but with the cost of much higher complexity. JPEG should be preferred, if the key aspects are the low cost and low power of the sensor node, as it is in our goal.

In this work we will compare retinex [19] and nonlinear stretching [20] image enhancement algorithms for different values of quality JPEG compression in presence of errors in the data stream due to noise in the channel. The measure used for the quality of image processing is the PSNR:

$$PSNR = 10\log_{10}\left[\frac{2^{2k}}{MSE}\right] \qquad (1.1)$$

where k is the number of bits representing each pixel, and MSE is the mean square error defined as

$$MSE = \frac{1}{n \cdot m}\sum_{i=1}^{n}\sum_{j=1}^{m}\left(I_1(i,j) - I_2(i,j)\right)^2 \qquad (1.2)$$

where n and m are the dimensions of the images and I1 and I2 are the images to be compared. Many black and white images have been used to verify the quality of the algorithm. The simulations have been performed in MATLAB.

Figure 1.2a reports an original black and white image in windows bitmap (bmp) format (320×240, black and white, 8 bit resolution, 76.800 bytes), that can be a good test for critical illumination conditions in video surveillance. Figure 1.2b reports the original image after a JPEG compression with 20% quality. Figure 1.2c reports the image after a jpeg compression with 20% quality, successive to the nonlinear stretching. Figure 1.2d reports the image after a jpeg compression with 20% quality successive to a retinex processing, the improvement in the quality of the dark areas is evident and the contrast is higher if compared to the nonlinear stretching algorithm. In all the cases the effect of the strong JPEG compression (Q20) after image enhancement does not reduce¡ the quality of the image in an evident way.

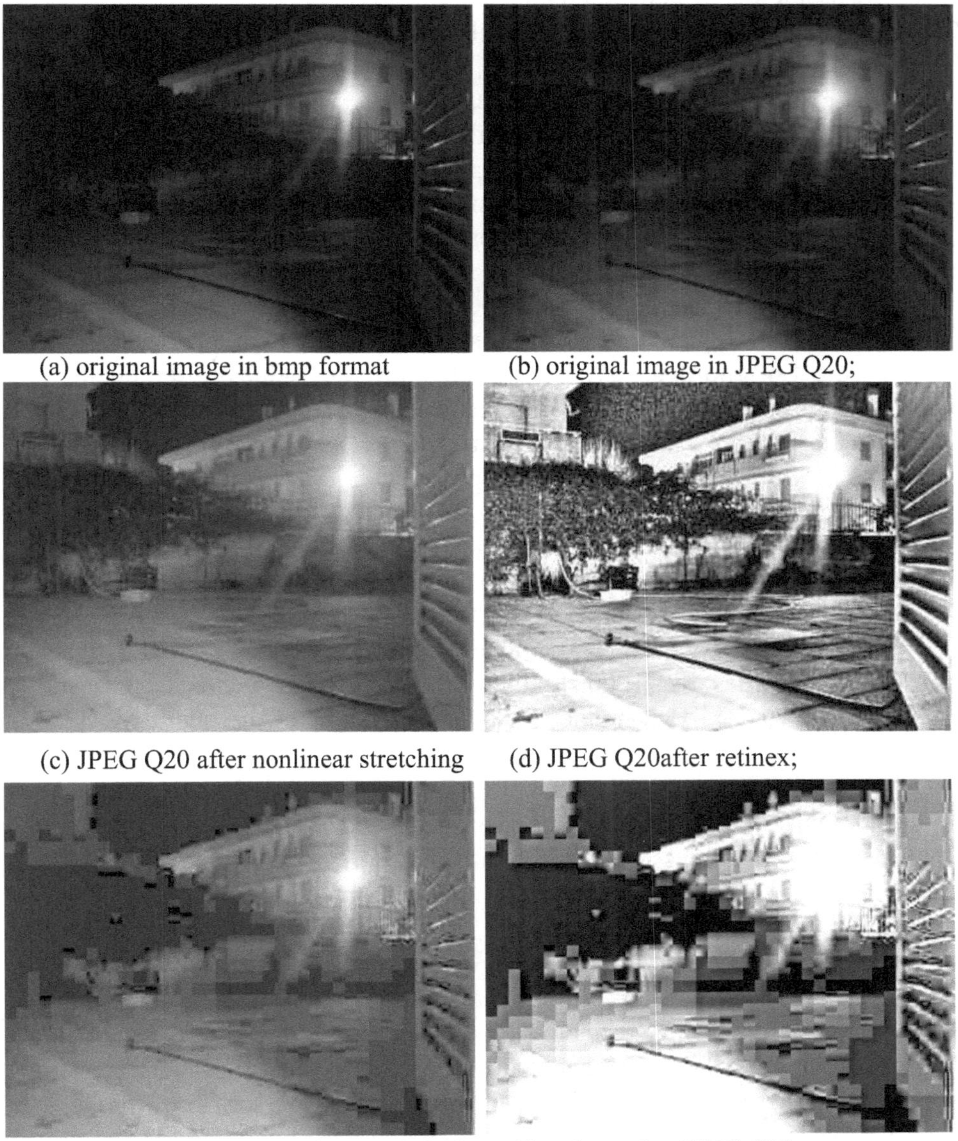

(a) original image in bmp format (b) original image in JPEG Q20;

(c) JPEG Q20 after nonlinear stretching (d) JPEG Q20after retinex;

(e) nonlinear stretching after JPEG Q20 (f) retinex after JPEG Q20.

Fig. 1.2 Image after different processing and compression tasks

Image enhancement (low complexity nonlinear stretching or retinex) must be performed to recognize the details of the image, indispensable for video surveillance. To verify the possibility to implement the image enhancement not in the sensor but in the receiver that has no energy limitations, we implemented the enhancement after JPEG compression, indispensable for reducing data rate and transmission power.

Figure 1.2e reports effects of nonlinear stretching applied to the image after a JPEG compression with 20% quality; and in Figure 1.2f retinex is applied to the image after a JPEG compression with 20% quality. The details of the image are no more visible. This means that the image enhancement must be performed in the image sensor before JPEG compression.

Figure 1.3 reports the compression ratio achieved applying JPEG to the original bmp image and after nonlinear stretching or retinex algorithms are applied, for different values of JPEG compression. The image enhancement, increasing the details, reduces the JPEG compression, as expected. But in the nonlinear stretching case, the compression is still very high.

Figure 1.4 reports the PSNR in dB for different values of the quality of JPEG compression and for the original, after nonlinear stretching or after retinex algorithms are applied. As an example the difference between the original (Figure 1.2a) and compressed image (Figure 1.2b) expressed in PSNR is 36 dB, while the PSNR between the retinex transformed image and compressed of retinex transformed image (Figure 1.2d) is 23 dB. This means that the image enhancement algorithm makes the JPEG compression critical.

Statistical simulation has been performed inserting in a random position errors in the images and the PSNR has been calculated. Figure 1.4 reports the average values over 1000 simulations of the PSNR when an error of only 1 bit is inserted in each image, for different values of the quality of JPEG compression.

Figure 1.5 reports the decrement in the PSNR (ΔPSNR) as a function of the number of bits of error inserted in each image. The dots represent the numerical simulations with different values of JPEG compressions. The dependence of ΔPSNR with JPEG quality is not relevant, as can be seen from Figs. 1.4 and 1.5. The statistical simulation shows that the degradation of the quality of the image does not strongly depend on the JPEG compression quality or on the size of the image but on the number of bits of error inserted. The image quality is still acceptable for video surveillance applications for a PSNR higher than 20dB corresponding to about 5 bits of error inserted in the image. From the simulations the following model of ΔPSN has been derived as a function of the n_e number of bits of error inserted in the image

$$\Delta PSNR = a + b\ln(n_e) \tag{1.3}$$

The model is shown in Fig. 1.5, the parameters *a* and *b* are obtained fitting the experimental data.

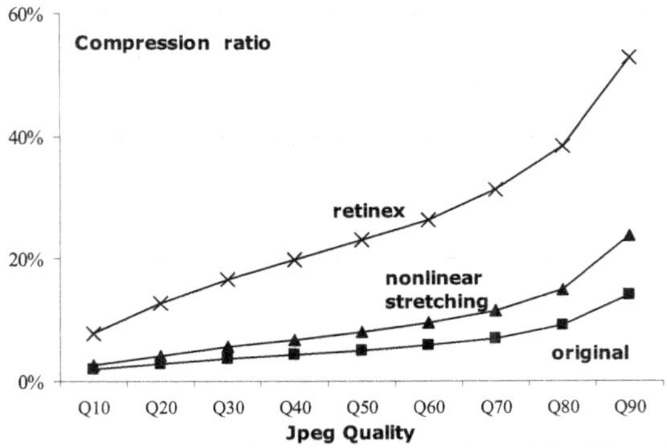

Fig. 1.3 Compression ratio with respect to the original bmp image

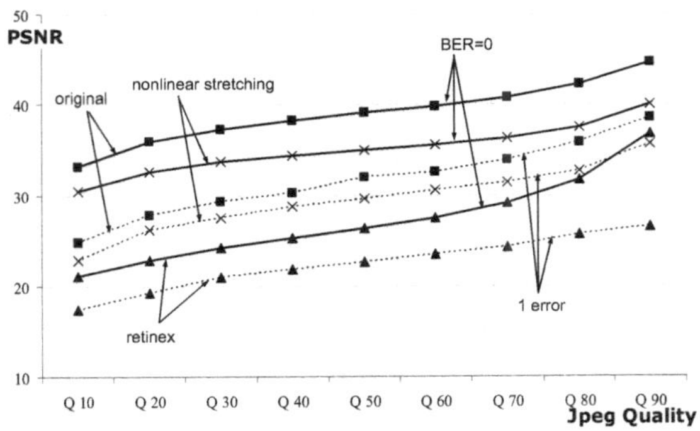

Fig. 1.4 PSNR in dB for different values of the JPEG quality, for the original, after nonlinear stretching or after retinex are applied, with an error of 1 bit and without errors

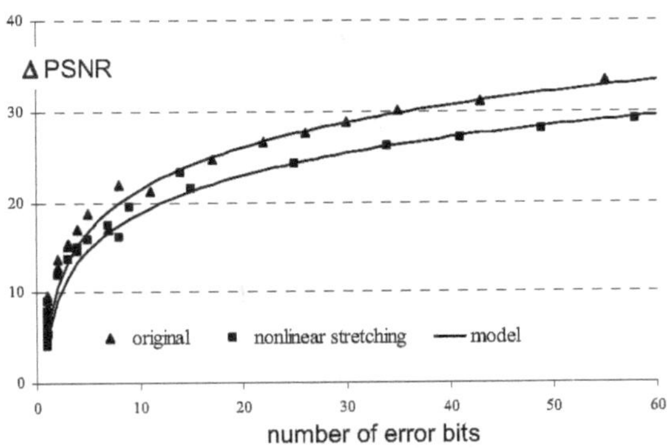

Fig. 1.5 Decrement in PSNR in dB for different values of errors in the image

1.3 Wireless Image Sensor Network Performance Analysis

A great number of standards have been defined for wireless communications, each one for different applications and with different characteristics.

Between them, widely used are the WiFi, Bluetooth, Home-RF, ZigBee, Wireless USB and Certified Wireless USB. In this work we will study the performance of Bluetooth [22] and ZigBee [23–25] for video surveillance applications, the topology considered is a star connection, as shown in Fig. 1.6 with images transmitted from the sensors to the central control station.

Bluetooth (IEEE 802.15.1) is mainly used for devices that have regular charge (e.g. mobile phones) and in applications like: hands-free audio, file transfer. The data rate is 1 Mb/s using a 1-MHz channel with an effective maximum bandwidth of 721kb/s. Bluetooth operates under a Time Division Duplex (TDD) polling scheme with a Frequency Hopping Spread Spectrum (FHSS) technique, with Gaussian Frequency Shift Keying (GFSK) modulation. Bluetooth allows many degrees of freedom to the designer, such as the choice of the different type of packets (DH1, DH3, DH5, DM1, DM3, DM5).

The ZigBee (IEEE 802.15.4) standard has been developed specifically for remote monitoring and control. ZigBee networks are designed to save the power of the slave nodes. For most of the time, a slave device is in deep-sleep mode and wakes up only for a fraction of time to confirm its presence in the network. The targets of ZigBee are low cost applications where the battery cannot be changed (battery life time of 1–2 years) with limited requirements of bandwidth. The maximum packet size is 133 bytes, but only 102 bytes can be used for data transmission. The supported nominal data rates are 250 kb/s, and the effective maximum date rate is about 190 kb/s. The ZigBee MAC layer uses the CSMA-CA algorithm to access the channel: each device tries to use the channel after a random delay. This algorithm yields high throughput and low latency for low duty cycle devices, like sensors and controls. Unfortunately the CSMA-CA algorithm reduces the effective maximum information rate to about 125 kbps in case of a single node that uses the channel. ZigBee is preferable for a network infrequently used and passing small data packets.

The Lower layers of Bluetooth and ZigBee have been implemented in SystemC [26–28]. Figure 1.7 shows the maximum network throughput as a function of the BER for Zigbee and Bluetooth, for different number of sensors (from 1 to 6) and for different types of Bluetooth packets. The data have been obtained with SystemC simulations [26, 27] and similar data are reported for the Bluetooth standard [29]. We considered a star topology with

1 master and n slaves, continuously transmitting images to the master. In a Bluetooth network the maximum throughput is independent of the number of slaves, therefore in Fig. 1.7 only one line can be seen for each type of Bluetooth packet.

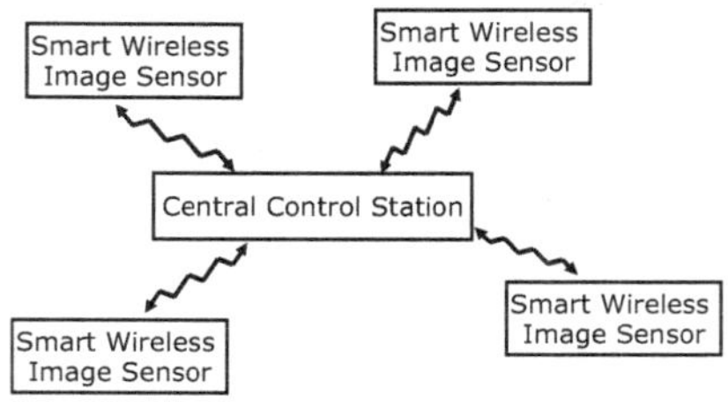

Fig. 1.6 Smart wireless sensor network topology

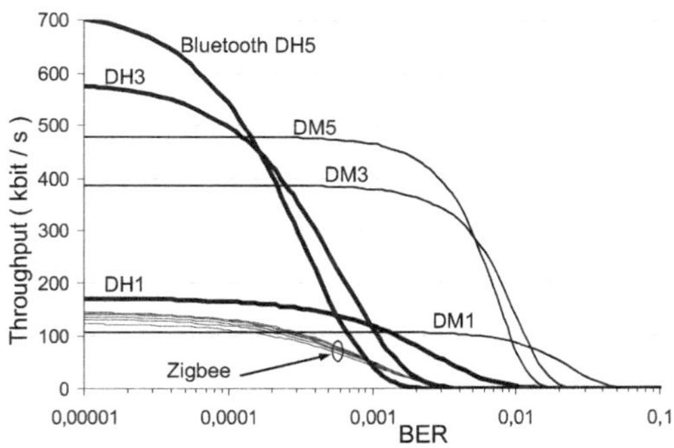

Fig. 1.7 Effective network throughput of Bluetooth and Zigbee as a function of the BER

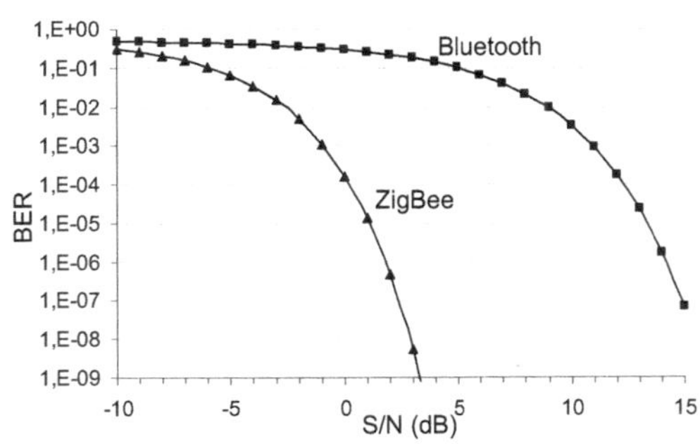

Fig. 1.8 BER as a function of signal to noise ratio for Bluetooth and Zigbee

In a Zigbee network the maximum throughput increases lightly with the number of sensors due to the CSMA-CA algorithm, therefore 6 curves are visible in Fig. 1.7 for the Zigbee network: the lower one corresponds to the network with 1 coordinator and 1 sensor, the upper one corresponds to the network with 1 coordinator and 6 sensors. The throughput of the Zigbee network is, in general, lower than Bluetooth throughput: the best type of Bluetooth packet depends on the BER.

The DM packets use 10/15 FEC with CRC allowing a correction of one bit of error for each 10 bit of data with the cost of reduction of throughput without errors; conversely the effective throughput is higher for DM packets with respect to DH packets for high values of BER. Bluetooth gives higher bandwidth, but the cost and the power dissipation for transmission and for the protocol processing is higher with respect to Zigbee.

Figure 1.8 shows the BER as a function of the signal to noise ratio for Bluetooth and Zigbee obtained from the 802.15.4 specifications [24,25], and it shows that Zigbee coding is more robust in presence of noise with respect to Bluetooth.

The number of images per second that can be transmitted by each sensor in the network depends on the size of the image, on the protocol, on the noise and on the number of devices in the network. As an example we consider the transmission of the image shown in Fig. 1.2c, that is the after nonlinear stretching and JPEG Q20 compression with a size of 25216 bits. In previous section, in particular in Figs. 1.3 and 1.4, it has been evidenced that nonlinear stretching with JPEG Q20 is a good compromise allowing good quality of image for video surveillance purposes, strong size reduction with respect to the 614400 bits of the original bmp image, and reduced decrement of PSNR in presence of noise.

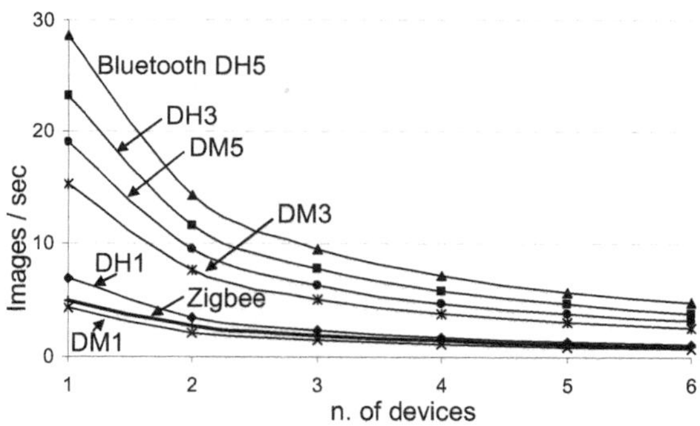

Fig. 1.9 Number of images per second that can be transmitted by each sensor as a function of the number of sensors in the network, for a Zigbee network and a Bluetooth network with different type of packets

Figure 1.9 shows the number of images (25216 bit size) per second that can be transmitted in absence of noise in the channel by each sensor as a function of the number of sensors in the network, for a Zigbee network and a Bluetooth network with different types of packets. In a Bluetooth network the throughput from each sensor to the master is simply the throughput achievable in the case of a network with only one sensor divided by the

number of sensors. In a Zigbee network, due to the randomness of the CSMA-CA channel access algorithm, the throughput from each sensor is higher than the throughput achievable in the case of a network with only one sensor divided by the number of sensors.
An analysis of the images sent in the networks in presence of noise can be obtained combining the analysis shown in Figs. 1.3 and 1.4 that considers the image enhancement algorithms and JPEG compression factor, with the analysis shown in Figs. 1.7 and 1.8, that considers the network throughput in presence of noise.

The BER in the transmission of a JPEG image degrades the quality of the image. In many cases the image cannot be recovered. This means that the image should be retransmitted. Figure 1.10 shows the number of images per second that can be send to the coordinator of a Zigbee network with 6 sensors as a function of the signal to noise ratio for the original bmp image not compressed, for the original image with JPEG Q20 and Q90 compression, for the image processed by retinex and successively compressed with JPEG Q20 and Q90, and finally for the image processed by nonlinear stretching and successively compressed with JPEG Q20 and Q90.

The number of images not compressed with JPEG (bmp in Fig. 1.10) per second is extremely low. The choice of a suitable image enhancement algorithm and the JPEG compression factor is important not only for the image quality but also for the image rate, and consequently for the power dissipation. The image rate can be improved 70 times with respect to bmp image, allowing the use of Zigbee for video surveillance applications. The Zigbee coding is efficient even in presence of noise, as shown in Fig. 1.8, therefore the image rate is degraded only with high values of noise.

Figures 1.11 and 1.12 show the same results of Fig. 1.10 but for a Bluetooth Network using DM1 and DH5 packets, respectively. The DM1 packets give the lower data rate with respect to the other Bluetooth packets; therefore the performances are worse than Zigbee even with low values of noise. Conversely the data rate of DH5 packets is strongly dependent on noise. Good results are evidenced using DH5 packets with low values of noise but the Zigbee performances are much more insensitive to noise.

1.4 Conclusions

Many simulations have been performed to investigate the effect on PSNR and on ZigBee and Bluetooth data rate of the parameters of JPEG compression for different image enhancement algorithms and in presence of noise in the channel.

In conclusion the retinex algorithm is more sensitive to the noise with respect to the nonlinear stretching that gives good image enhancement even with JPEG compression and in presence of noise.

Both ZigBee and Bluetooth protocols are suitable for video surveillance applications; Zigbee seems to be more robust to noise with respect to Bluetooth. Conversely the frequency hopping technique used by Bluetooth makes the network more insensitive to the interference with other 2.4GHz networks.

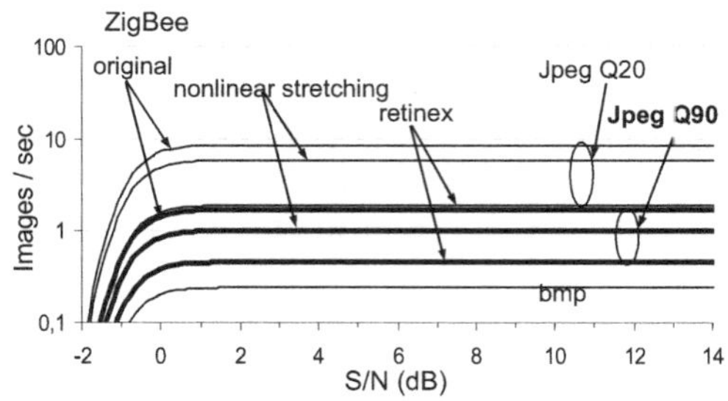

Fig. 1.10 Images / s that can be send to the coordinator of a Zigbee network with 6 sensors as a function of S/N for different image processing in the sensor

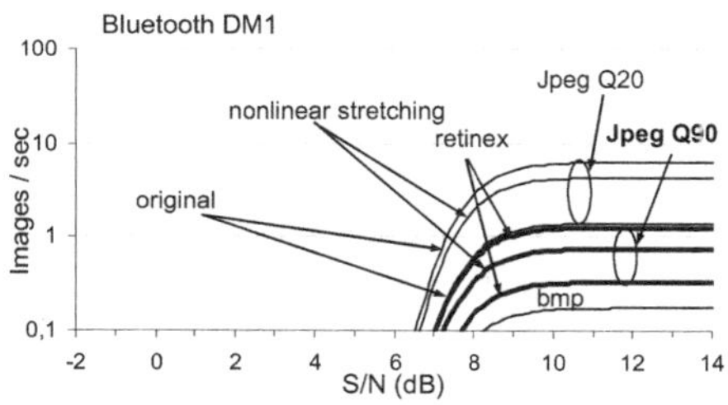

Fig. 1.11 Images / s that can be send to the master of a Bluetooth network using DM1 packets as a function of S/N for different image processing in the sensor

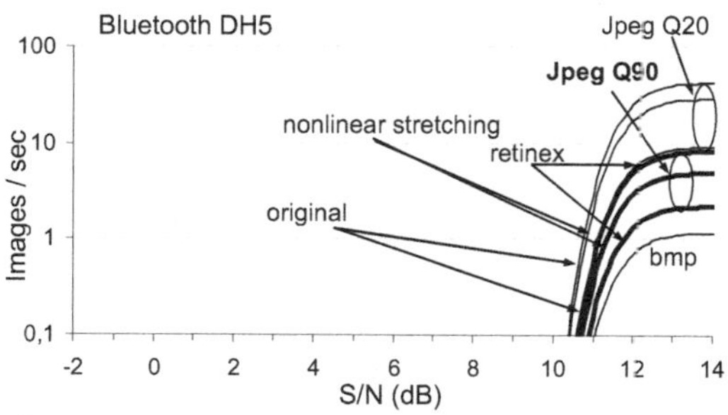

Fig. 1.12 Images / s that can be send to the master of a Bluetooth network using DH5 packets as a function of S/N for different image processing in the sensor

References

1. L.-W. Lai, C.-H. Lai, Y.-C. King, "A Novel Logarithmic Response CMOS Image Sensor with High Output Voltage Swing and In-Pixel Fixed-Pattern Noise Reduction", IEEE Sensors Journal, 4(1), 122–126, Feb. 2004.
2. I.F. Akyildiz, W. Su, Y. Sankarasubramaniam, E. Capirci, "Wireless Sensor Networks: A Survey", Computer Networks 38, 393–422, 2002.
3. C.-Y. Chong; S.P. Kumar, "Sensor networks: evolution, opportunities, and challenges", Proceedings of the IEEE, 91(8), Aug. 2003.
4. A.P. Chandrakasan, R. Min, M. Bhardwaj, S.H. Cho, A. Wang, Power Aware Wireless Microsensor Systems, keynote Paper ESSCIRC, Florence, Italy, September 2002.
5. K. Sohrabi, J. Gao, V. Ailawadhi, G.J. Pottie, Protocols for Self-Organization of a Wireless Sensor Network, IEEE Personal Communications, Vol. 7, No. 5, Oct. 2000.
6. W. Ye, J. Heidemann, D. Estrin, An Energy-Efficient MAC Protocol for Wireless Sensor Networks, In Proceedings of the 21st International Annual Joint Conference of the IEEE Computer and Communications Societies (INFOCOM 2002), New York, NY, USA, June, 2002.
7. L. Hampapur, J. Brown, A. Connell, N. Ekin, M. Haas, H. Lu, S. Merkl, S. Pankanti, "Smart Video Surveillance: Exploring the Concept of Multiscale Spatiotemporal Tracking," IEEE Signal Processing Mag., 22(2), 38–51, Mar. 2005.
8. G.L. Foresti, C. Micheloni, L. Snidaro, P. Remagnino, T. Ellis, "Active Video-Based Surveillance System: The Low-Level Image and Video Processing Techniques Needed for Implementation," IEEE Signal Processing Mag., vol. 22, no. 2, pp. 25–37, Mar. 2005.
9. R. Kleihorst, A. Abbo, V. Choudhary, B. Schueler, "Design Challenges for Power Consumption in Mobile Smart Cameras," in Proc. COGnitive Systems with Interactive Sensors (COGIS 2006), Mar. 2006.
10. M. Rahimi, R. Baer, O.I. Iroezi, J.C. Garcia, J. Warrior, D. Estrin, M. Srivastava, "Cyclops: In Situ Image Sensing and Interpretation in Wireless Sensor Networks," in Proc. 3rd International Conference on Embedded Networked Sensor Systems (SenSys 2005), Nov. 2005, pp. 192–204.
11. F. Dias Real de Oliveira, P. Chalimbaud, F. Berry, J. Serot, F. Marmoiton, "Embedded Early Vision Systems: Implementation Proposal and Hardware Architecture," in Proc. COGnitive systems with Interactive Sensors (COGIS 2006), Mar. 2006.
12. I. Downes, L. Baghaei Rad, H. Aghajan, "Development of a Mote for Wireless Image Sensor Networks," in Proc. COGnitive Systems with Interactive Sensors (COGIS 2006), Mar. 2006.
13. Z.-Y. Cao, Z.-Z. Ji, M.-Z. Hu, "An Image Sensor Node for Wireless Sensor Networks," in Proc. International Conference on Information Technology: Coding and Computing (ITCC 2005), 2, 740–745, Apr. 2005.
14. R. Kleihorst, B. Schueler, A. Danilin, M. Heijligers, "Smart Camera Mote with High Performance Vision System," ACM SenSys 2006 Workshop on Distributed Smart Cameras (DSC 2006), Oct. 2006.
15. M. Bramberger, A. Doblander, A. Maier, B. Rinner, H. Schwabach, "Distributed Embedded Smart Cameras for Surveillance Applications," in IEEE Computer Mag., 39(2), 68–75, Feb. 2006.
16. S. Hengstler and H. Aghajan, "WiSNAP: A Wireless Image Sensor Network Application Platform", 2nd Int. Conf. on Testbeds and Research Infrastructures for the Development of Networks and Communities (TridentCom), March 2006.

17. S. Hengstler, D. Prashanth, S. Fong, H. Aghajan, "MeshEye: A Hybrid-Resolution Smart Camera Mote for Applications in Distributed Intelligent Surveillance", Information Processing in Sensor Networks (IPSN-SPOTS), April 2007.
18. L. Brown, J. Connell, A. Hampapur, M. Lu, A. Senior, C.-F. Shu, Y. Tian, "IBM Smart Surveillance System (S3): A Open And Extensible Framework For Event BasedSurveillance", Proc. of the Int. Conf. on Advanced Video- and Signal-based Surveillance September 2005.
19. E.H. Land, "Recent Advances in Retinex Theory", Vision Research, 26(1), 7–21, 1986.
20. T. Balercia, A. Zitti, H. Francesconi, S. Orcioni, M. Conti, "FPGA implementations of a simplified Retinex image processing algorithm", Proc. of IEEE ICECS '06 Int. Conf. on Electronics, Circuits and Systems, Nice, France, pp.176–9, December 10–13 2006.
21. G. Pekhteryev, Z. Sahinoglu,P. Orlik, G. Bhatti, "Image transmission over IEEE 802.15.4 and ZigBee Networks", Proc. of IEEE ISCAS 2005, Kobe, Japan. pp.3539–3542.
22. Bluetooth SIG, "Specification of the Bluetooth system version 1.1," Feb 2001.
23. "ZigBee White Paper" and "ZigBee Specifications" in www.zigbee.org
24. IEEE 802.15.4TMStd.– 2003 – Wireless Medium Access Control and Physical Layer Specifications for Low-Rate Personal Area Networks (WPANs).
25. IEEE 802.15.4TM Std.–2006– Revision of IEEE 802.15.4TM Std. – 2003.
26. M. Caldari, M. Conti, P. Crippa, G. Marozzi, F. Di Gennaro, S. Orcioni, C. Turchetti, "SystemC Modeling of a Bluetooth Transceiver: Dynamic Management of Packet Type in a Noisy Channel" Proc of DATE 2003, Munchen, pp.214–219.
27. M. Conti, D. Moretti, "System Level Analysis of the Bluetooth Standard", Proc of DATE 2005, München, 3, 118–123.
28. A. Mignogna, M. Conti, M. D'Angelo, M. Baleani, A. Ferrari, "Transaction Level Modeling and Performance Analysis in SystemC of IEEE 802.15.4 Wireless Standard" DSD'2008, Proc. of Euromicro Conf. on Digital System Design, 3–5 September, 2008, Parma, Italy.
29. T.Y. Chui, F. Thaler, W.G. Scanlon, "A Novel Channel Modeling Technique for Performance Analysis of Bluetooth Baseband Packets", ICC 2002. IEEE International Conf. on Communications 2002, 1, 308–312, 28 April-2 May 2002.

Chapter 2

Policy Management Architecture for Multimedia Services in a Multi-Provider Scenario

Mario Ibáñez, Natividad Martínez Madrid and Ralf Seepold
Universidad Carlos III de Madrid, 28911 Leganés (Madrid), Spain,
mario.ibanez@uc3m.es, natividad.martínez@uc3m.es, ralf.seepold@uc3m.es

Abstract The management of multimedia services integrated into a service platform has been taken as a starting point for the development of a new model that supports virtualization of the platform concept. The virtualization of provider dependent configuration frameworks is introduced to isolate different service providers during the installation and maintenance of new services contracted by the end user. In addition, the virtualization has to be enhanced with a system that provides a management of common resources and the way of accessing to them. That is made with a set of policy management rules and procedures that are described in this paper. The model has been evaluated in a multi-provider case study with focus on multimedia data management.

Keywords Service platforms, Residential gateways, Policy management.

2.1 Introduction

Multimedia services are arriving to our homes. Each month, new offers of Video on Demand, TV over IP, etc. are being including in the service catalogue of the telecom operators. It is important to enable a proper infrastructure to the deployment of these services in the home. Therefore, this infrastructure needs also to take into account that multiple multimedia services arrive concurrently to the home network and it is needed to maintain a certain *Quality of Service* (QoS).

The coexistence of multiple services demands a platform that is dynamically managing the life cycle of all services. The OSGi platform

N. Martínez Madrid, R.E.D. Seepold (eds.), *Intelligent Technical Systems*, Lecture Notes in Electrical Engineering 38, DOI 978-1-4020-9823-9_2,

provides a framework for the execution and management of services.[a] This service platform allows installing and uninstalling different services in the same device, the Residential Gateway (RGW). The RGW is placed at the border of the home network and manages data flows that enter or leave the home environment. Those flows correspond among others to multimedia services that are managed by the applications installed in the OSGi platform.

In an open market, all services that the user has contracted can be served by more than one service provider. This fact draws a different scenario in which different administrators have to manage the RGW. Shared management of a system is a complex task, with several consequences for security. Different solutions decide to avoid or prohibit by definition more than one user management of the RGW, limiting in this way the business models and functionality. On the contrary, this chapter presents a solution that allows to provide this kind of behavior. The solution consists in the creation of new virtual platforms placed over the service platform to provide isolation to the management of each administrator. That means that every administrator of the service platform can modify its applications in the RGW without the intervention of any other administrator.

In Section 2.2, the state of the art is presented and the OSGi technology is introduced. In Section 2.3, the working scenario is presented and roles are described in detail. Section 2.4 presents the software architecture of the solution and Section 2.5 summarizes the results obtained. Finally, Sections 2.6 and 2.7 summarize the chapter and future work is presented.

2.2 State of the Art

The proposed management architecture is based on the OSGi service platform which is used to manage services and hosts the management and the virtualization systems of the platform. The next sections present an analysis of related research work and a summary the relevant characteristics of OSGi.

2.2.1 Related Work

The work presented in this paper concentrates on the development of a service platform for a RGW oriented to multimedia services. A basic model of a service platform has been presented by Hofrichter [1] and Waring [2]; a

[a] http://www.osgi.org

more complete and also updated definition is given by the Home Gateway Initiative (HGI) [3]. Since service provisioning is relevant also for a multi-provider scenario, a simple scenario based on OSGi is presented by Dueñas [4]. It describes a scenario with different users involved in the management and deployment of services in the RGW. Furthermore, two main users are defined; the operator and the service provider. The operator is in charge of managing the RGW and the service provider is in charge of providing services. However, the service provider cannot manage services in the RGW. This approach lacks support for multiple users managing the platform at the same time but it presents a first approach to the management of RGW.

A different management approach is presented by Cho [5]. In this model, the management of services is tackled as a policy access problem and it is solved using the Role-Based Access Control (RBAC) model. This model defines users, sessions, roles, permission and the associations between those entities. Surrounding this model, an infrastructure and a protocol is created to provide permission to service providers for managing services in the RGW. The model includes an access provider who defines access rights for the different service providers. The end users contract services, the model notifies changes and enables the required permissions. This model is also based on the assumption that only one of the different types of users has the full control and thus takes the final decisions about the RGW configuration. Again, this is not a multi-manager approach.

Finally, there is another relevant approach [6] that proposes different users being members of the RGW management at the same time. In addition, this paper proposes a concept of virtual platforms. Isolation is the basic mechanism to separate different users and thus to avoid interaction. There are only two different managers of the RGW, the operator controlling the service platform and authorizes the service providers, while the service providers actuate as users that can manage the virtual platforms. This approach does not support an end user managing own bundles.

The solution shown in this chapter is based on the concept of virtualization as the way of isolating different managers (access providers, service providers, end users and system vendors). These managers can simultaneously access to the RGW and configure its behavior modifying the basic parameters of the router or installing new services. This is possible due to the design of a new virtualization bundle installed in the OSGi platform. This bundle provides the needed abstraction over the router and OSGi platforms while providing isolation to the different virtual platforms. This model is based on classical virtualization techniques.

2.2.2 The Open Service Gateway Initiative

OSGi [7] has been selected as the platform for the services in the RGW. The OSGi specifications define a standardized, component oriented, computing environment for networked services. Adding an OSGi Service Platform to a networked device (embedded as well as servers) enables the capability to manage the life cycle of the software components in the device from anywhere in the network.

Software components can be installed, updated, or removed on the fly without having to disrupt the operation of the device. The software components or applications are called bundles. When activated, bundles may register services to provide services to other bundles in the framework

OSGi specifications also define a way for the remote management of the platform based on two components: the Remote Manager (in the mangers side) and the Management Agent (in the RGW). There is no communication protocol predefined.

The Remote Manager is a tool provided for the operator to control the service platform. By using it, the operator can access all the functions determined in the remote management interface. However, this manager is not clearly defined in the standard, leaving a concrete implementations open, i.e. to adapt to the specific properties of the management protocols used.

The Management Agent is the system's entry point for the Remote Manager. The Management Agent is a bundle that is called a Management Bundle. It owns permission for the administration of the platform, so it can manage the life cycle of the bundles in the RGW. The platform can have more than one Management Agent, each one communicating with a different remote manager and using different management protocols.

2.2.3 Virtualization

A solution based on virtualization has been developed to isolate the different management agents of each service provider. As a result, each virtual platform is managed by only one service provider. As the RGW can host several virtual platforms, it will have several service providers. In the following, the basics of virtualization are commented by Smith [8].

In general, virtualization is based in the concept of a machine being constructed over different abstractions and interfaces. An abstraction is used to ignore or simplify the lower level using a well defined interface. Virtualization is a way of relaxing the constraints and increasing the flexibility. It also constructs and isomorphism that maps a virtual system on an underlying system and differs from an abstraction because the virtualization does not necessarily hide details.

The products of the virtualization are the Virtual Machines (VM) that are layers of software on top of the real machine to support the desired virtual machine's architecture. The VM can be used to replicate systems, providing independency and isolation between each replication, or to emulate the support of cross platform software, as well as to describe architectures that do not correspond to a real machine.

A first action to be done in the design of the virtualization system is to identify the interface for the virtual machine. Once the interface has been indentified, two steps have to be taken to construct a virtual machine: Firstly, a research about how to map resources of the underlying machine, and secondly, the development of the actions which allow the communication between the real machine and the virtual machine.

2.3 Working Environment

To understand the work presented, it is necessary to know the environment in which this work has been carried out. Then, the device and networks involved will be defined. After that, the different roles that the user of the RGW can take will be presented, and finally, a business model is sketched out.

2.3.1 Scenario Description

Figure 2.1 shows the key role of the RGW because it is located between the Home Network and the Access Network. Therefore, it is an ideal place to install a service platform that manages home services like for example multimedia services. Moreover, the RGW provides router functionality that connects to different networks.

As mentioned before, the OSGi framework, provides and execution environment to host application services. Due to the different services and data flows that the RGW must handle, it is necessary to install a management service allowing an easy access to the configuration data of the device. The management service provides access to the RGW configuration, like for example the data flow definition in the router or the access to the life cycle of services using OSGi.

The Home Network is full of different equipment for different issues some of them are related to multimedia like TVs. These devices capable of playing or serving multimedia content require an infrastructure to share multimedia content which sometimes will arrive from outside home. The management of those multimedia flows can be implemented with different protocols, for example with UPnP installed as a service in the RGW.

The access network is managed by the access provider. This network is in charge of providing multimedia content with QoS, like it is provided by the service providers. The access network is also the entrance point for flows from Internet.

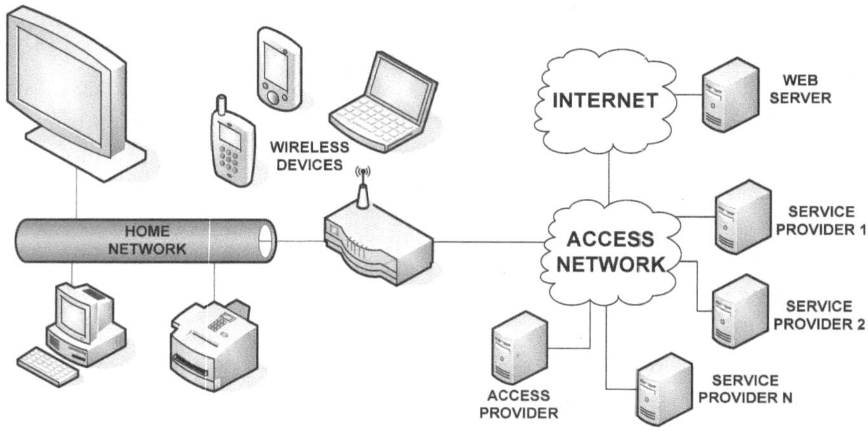

Fig. 2.1 RGW environment

In a basic scenario, the management of the access networks as well as the management of the RGWs of the end users is done by the access provider. But in a more complex scenario the service providers can manage directly the RGWs according to the services offered, while the access provider only manage the traffic in the network to maintain the QoS.

2.3.2 Manager Roles

Four different managers, required for the setup of the scenario, are presented in Fig. 2.2. The access provider is in charge of ensuring the QoS in the access network. It also provides a way of connecting an end user to a service provider. So, in relation with the RGW, it can adapt the configuration of the QoS in the RGW to the needs in the network, and thus manage their applications in the platform. Basically, this application will be the one in charge of the management of the QoS.

The service provider will provide services to the end users. That means that it will be necessary to modify the QoS parameters for its own services and of course to manage its own services installed in the RGW. The service provider is not present in the initial configuration of the RGW but this role is added in the moment when a service is enabled in the RGW.

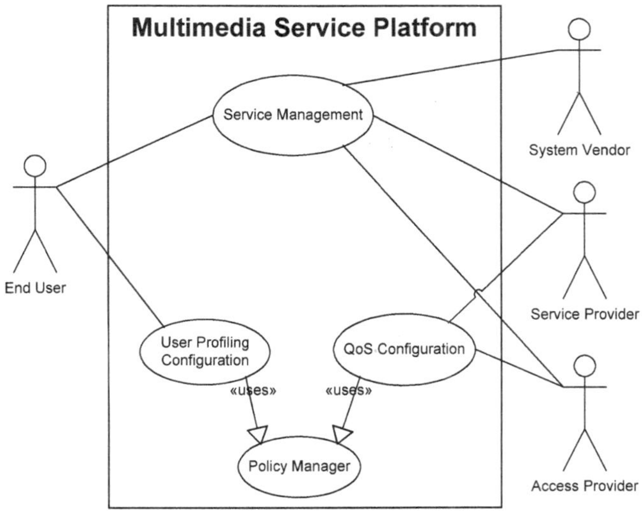

Fig. 2.2 Roles in the RGW

The system vendor is in charge of providing a service for updating the firmware of the machine. Also this role needs to access to the management of that service.

Finally, there is the end user. This role is mainly in charge of contracting new services which are later installed by the services provider in the RGW. Therefore, it needs to access to, at least, a basic configuration of its applications. In a more broad sense, it is possible to find different end users with different objectives and applications that can be managed independently. For example a leisure user that hires video on demand services and a professional user that hires bandwidth to make videoconferences with a certain QoS. This differentiation of roles forces an establishing of a user profiling configuration to define how the RGW will behave.

All data related to the configuration of QoS or profiles will be managed by a service which determines the policy to be followed by the RGW.

2.3.3 Business Model

Once the scenario has been setup and the different users have been presented, a model is required that defines how relationships between them are established.

We assume that the end user buys a RGW from a manufacturer or from an access provider. In case the RGW has been bought from a manufacturer it may cover a maintenance contract. The manufacturer needs to have a bundle to update the basic installation of the device. In the second case, the maintenance of the firmware can be done by the provider. In both cases, the access provider needs permission to manage the RGW.

In case other relationships are created between the end user and the access provider, the access provider has to install a bundle in the service platform to manage the QoS parameters.

A third relationship is established between service and access providers: Both agree that the service provider can use the access network to deliver services. This relation includes an agreement to maintain the QoS in the network.

The last relation is established between the end user and the service provider in order to have a service in the home. This relation also implies the installation of services in the RGW and the management of them by service providers.

2.4 Policy Management Architecture

The management has evolved from the management of one operator to multiple managers. Next two sections present the evolution from the mono-provider management to the multi-provider management with virtualization. Finally, details for this specific policy management are presented

2.4.1 Mono-Provider Management

The mono-provider management scenario (cf. Fig. 2.3) presents three main devices in the access network; the RGW, the Auto-Configuration Server (ACS) and the Service Aggregator. The ACS stores the data for configuring the different services and installs them in the RGW as they are needed. The service aggregator provides a unique place from where it is possible to deploy services, although, they are owned by different service providers.

The solution implemented for the remote and automatic configuration of the RGW is based on TR-069 protocol [9] from DSL. This protocol defines a set of parameters to be exchanged between an (ACS) and the RGW to provide the configuration. Moreover, the solution is modular and compound of three bundles that provide the configuration which are briefly described in Ibáñez [10].

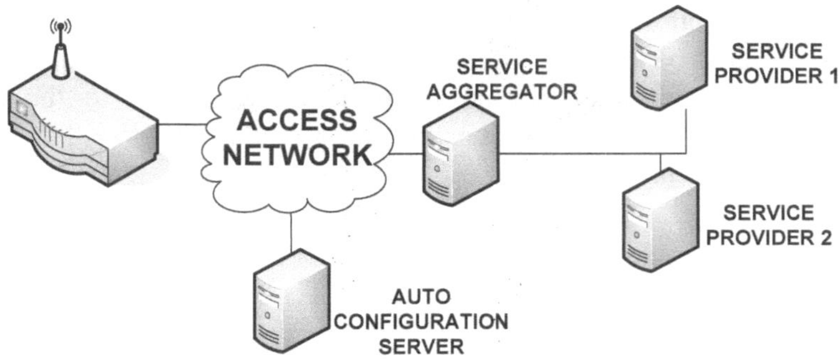

Fig. 2.3 Access Network in a mono-provider scenario

2.4.2 Multi-Provider Management

The objective of virtualization is to add mechanisms to the OSGi platform that allow having different instances of the platform that can be managed independently. However, having all virtualizations inside the same OSGi platform allows the use of common services as presented in Fig. 2.4.

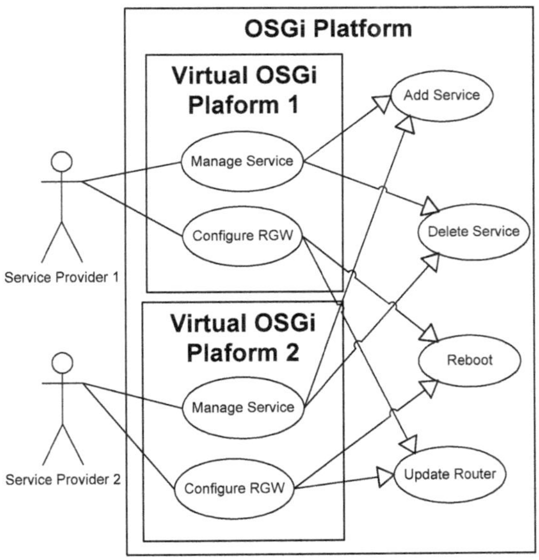

Fig. 2.4 Virtualization use case

This figure shows a scenario where two service providers want to manage the service platform. Both users of the RGW will manage the lifecycle of the bundles installed that means since there is not root in the gateway, one provider can uninstall bundles of the other provider. The new model solves this problem providing a kind of virtualization. Each virtual platform created presents to the user an interface to manage the/his whole system as it would be the only authorized user.

The objective of virtualization is to add mechanisms to the OSGi platform that allow having different instances of the platform that can be managed independently.

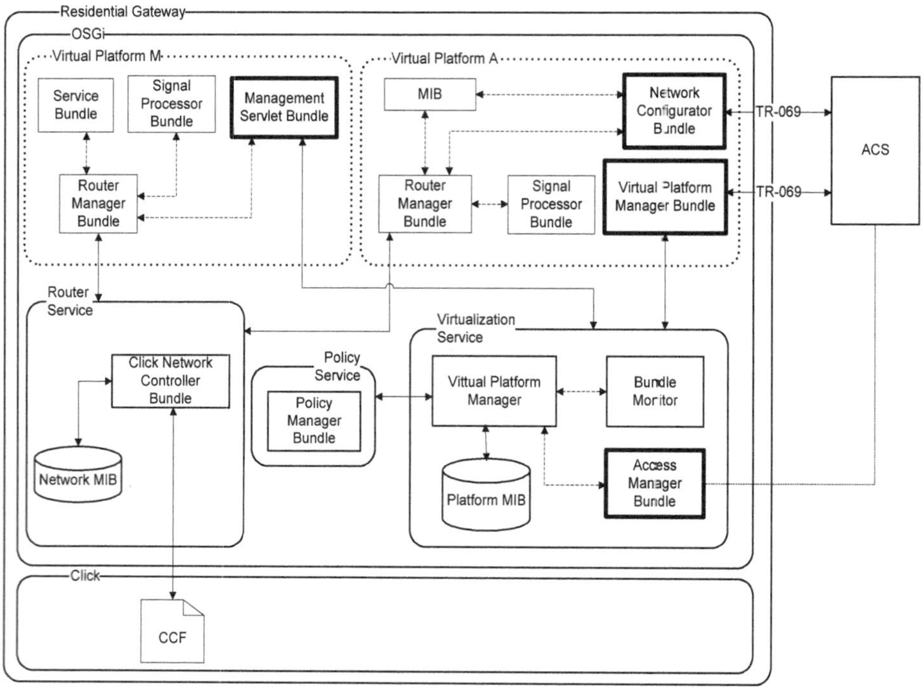

Fig. 2.5 Multi-provider functional block diagram in the RGW

The "Virtualization service" is a set of bundles that manages all virtual platforms. It is in charge of creating new virtual platforms, managing the registration of local and global services (Virtual Platform Manager), monitoring (Bundle Monitor) the use of resources by bundles and providing a generic access agent (Access Manager Bundle) to allow external entities to create new virtual platforms. Permissions for creating virtual platforms and installed services are checked and managed dynamically in the "Policy Manager Bundle". This bundle is usually managed by the access provider.

In addition to the infrastructure for virtualization there are *virtual platforms*. Each one has its own bundles and services and its own manager agent. In case of Fig. 2.5 there are two virtual platforms, one of them corresponds to the one presented in the mono-provider approach and uses two different Management Agents: “Network Configurator Bundle” and “Virtual Platform Manager Bundle”. The Virtual Platform A is used to manage the router automatically with TR-069 protocol. The Virtual Platform M platform contains only one Management Agent, the “Management Servlet Bundle” which is a servlet for manual management of both, router and services. I.e. the bundles and services of the virtual platform and those which are not in the virtual platform but which have been installed by the provider.

This architecture implies four different kinds of users with different roles: One is the end user who will be enclosed in a virtual platform via a user interface. Furthermore, there is the system vendor user who will access to the RGW in order to upgrade the firmware. A third user is the access provider who defines the policies to access the RGW. Finally, the service providers only access to their virtual platform to manage their own services.

2.4.3 Policy Management

Not all aspects of the management of the RGW by multiple users are covered by virtualization as is the case of the access to the services of a user by another user. In this case it is needed to establish a way of manage this issues out of the virtualization. Like other configuration issues it must have dynamic character due to dynamically added and deleted services and service provider. There are different aspects to be considered as management policies as the creation of virtual platforms.

When a service provider or an end user needs to create a virtual platform they have to be entitled to create it. In case of end users, a preconfigured virtual platform is provided by the RGW. This virtual platform owns bundles which provides and interface for the management of the RGW by the user. The task that this interface allows to do includes the creation of new end users and new virtual platforms for those users; it has the role of an administrator. In the case of service providers, the access provider has to allow the creation of the virtual platform as is shown in Fig. 2.6. The user hires a service from a service provider, then, this one sends a request for access to the access provider because he wants to configure the RGW and to allow the service provider to create a virtual platform. After that, the service provider can create a virtual platform installing his management. Via his management bundle he will install the service and he can configure the router.

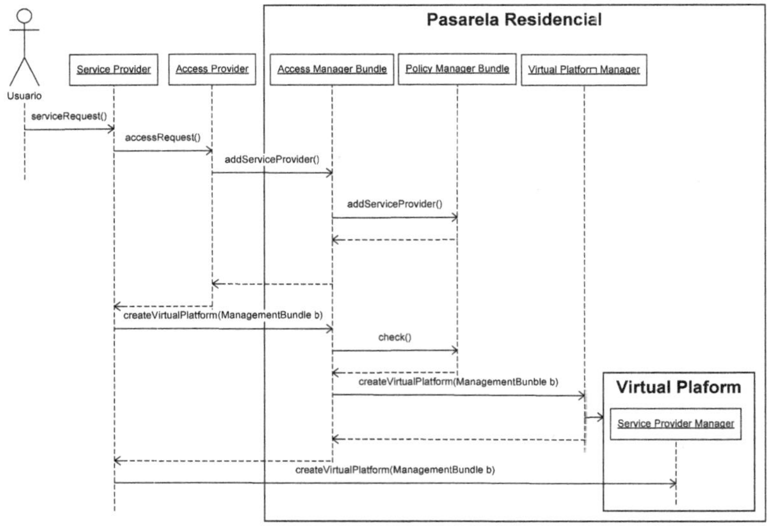

Fig. 2.6 Sequence diagram for creating a virtual platform for a service provider

The control of consuming resources is another issue required and thus policies are needed. The bundle monitor checks the use of resources of the different bundles with the possibility of grouping it into virtual platform. It is possible to limit CPU resources, memory or bandwidth per virtual platform or bundle so if this limit is exceeded, a corrective action can be performed. These actions are: stop the bundle that is abusing of the resource, stop one of the bundles of a virtual platform or do nothing if the use the resources are not in a critical situation. The manger of the virtual platform is in charge of defining which bundle is candidate to be stopped in case the virtual platform exceeds the limit. The access provider rules the limits for the bandwidth. The end user defines all limits for the use of CPU or memory although these limits should be default ones.

A final issue to be managed by the Policy Manager Bundle is profiles of different end users and their rights to install services. This is configured by the end user *root*.

2.5 Results

A demo scenario has been developed to check the viability of the proposed model. The RGW has been implemented in an embedded PC with a CPU VIA C3 533MHz. The PC is running under Linux Ubuntu 6.06. J2SE 1.5 has been selected for the installation and development of the bundles. For the OSGi installation the distribution Oscar 1.5 [11] has been used and the version 1.5 of the Click [12] software router has been selected. Also, two multimedia servers have been created to simulate different service providers.

The demo scenario has been implemented with all devices described. The access network has been simulated over an Ethernet network with injected traffic. In the home network scenario two laptops have been used as players of the multimedia content.

The complete functionality has been tested in a small case study. The results obtained confirm the functionality of the model and the implementation. It was possible to manage data flows and to change the QoS for the simulated multimedia services.

2.6 Conclusions

This chapter presents a model for the management of a RGW in a multi-provider scenario. It is based on the virtualization of the service platform that runs in the RGW. This virtualization is enhanced by a management of policies and aspects as common resources management and access policies.

It provides service and device integration independently from any technology, and furthermore, it interconnects the home network to the access network. The model proved to provide the mentioned capabilities without imposing a restriction on the number of remote management units or service providers. The feature to provide policy management of the different resources has been realized via new services that have been added to the RGW platform. These services are sharing several infrastructural framework parts in order to keep the overall overhead low.

2.7 Future Work

One aspect of the future work will be dedicated to further modularize the virtual platform architecture and thus to provide a pool of features for a multi-provider concept derived from the current needs. A second part of the work is to integrate the virtual platform prototype into a gateway provided by network operators. It is assumed that several services can be omitted since not each RGW will need to provide the full set of functionalities contained in the proposed model. Finally, as a more ambitious way, it is planned to provide the needed functionality for a device connected to different access networks.

Acknowledgment This work has been partly supported by the Multi Service Access Everywhere (IST-Program; MUSE-IST 026442) [13] project that is an Information Society Technologies (IST) priority of The Sixth Framework Program (FP6) of the European research activities and by the PLANETS project [14] (MEDEA+ A306 funded by the Spanish Ministry of Industry; FIT-330220-2005-111), and finally, by the MEC I+D project InCare-FAMENET. Ref: TSI2006-13390-C02-02.

References

1. K. Hofrichter, "The Residential Gateway as Service Platform". ICCE, International Conference on Consumer Electronics. Jun. 2001. ISBN: 0-7803-6622-0
2. D. Waring, "Residential Gateway Architecture and Network Operations," International Organization for Standardization (ISO) and International Electrotechnical Commission (IEC) document: JTC 1/SC 25/WG 1 N 848, May 1999
3. Home Gateway Initiative,http://www.homegatewayinitiative.com, June 2008
4. J.C. Dueñas, J.L. Ruiz and M. Santillán, "An End-to-End Service Provisioning Scenario for the Residential Environment", IEEE Communications Magazine, vol. 43, no. 9. Sep. 2005
5. E. Cho, C. Moon, D. Park and D Baik, "An Effective Policy Management Framework Using RBAC model for Service Platform based on Components", Software Engineering Research, Management and Applications, 2006. Fourth International Conference on, (09–11), 281–288. Aug. 2006. ISBN: 0-76952-656-X
6. Y. Royon, S. Frénot and F. Le Mouel, "Virtualization of Service Gateways in Multi-provider Environments", 9th International SIGSOFT Symposium on Component-Based Software Engineering (CBSE06), Stockholm (Sweden), Jun. 2006
7. Open Service Gateway Initiative (OSGi) Alliance. Jun. 2007. http://www.osgi.org
8. J.E. Smith and R. Nair, "Virtual Machines: versatile platforms for systems and processes", Elsevier, June 2005. ISBN: 1-55860-910-5
9. DSL Forum, "TR-069: CPE WAN Management Protocol", May 2004. http://www.dslforum.org/aboutdsl/Technical_Reports/TR-069.pdf
10. M. Ibañez, N. Martínez Madrid and R. Seepold, "An OSGI-based Model for Remote Management of Residential Gateways", LNCS: 10th Asia-Pacific Network Operations and Management Symposium, APNOMS 2007, Vol. 4773, Springer LNCS, ISSN: 0302-9743, 2007
11. Oscar OSGi platform, Jun. 2007. http://forge.objectweb.org/projects/oscar/
12. The Click router, Jun. 2007. http://www.pdos.lcs.mit.edu/click/
13. Multi Service Access Everywhere (MUSE) European Project, MUSE-IST 026442. Jun. 2008. http://www.ist-muse.org/
14. Platforms for Networked Service Delivery (PLANETS), MEDEA+ project A-121, mostly financed by the Spanish Ministry of Industry under project Num. FIT-330220-2005-111, Jun. 2008. http://www.medeaplus.org

Chapter 3

Embedding Multi-Task Address-Event-Representation Computation

Carlos Luján-Martínez, Alejandro Linares-Barranco, Gabriel Jiménez and Antón Civit
Depatment Arquitectura y Tecnología de Computadores, Universidad de Sevilla, Sevilla, SPAIN, cdlujan@atc.us.es, alinares@atc.us.es, gaji@atc.us.es, civit@atc.us.es

Abstract Address-Event-Representation, AER, is a communication protocol that is intended to transfer neuronal spikes between bioinspired chips. There are several AER tools to help to develop and test AER based systems, which may consist of a hierarchical structure with several chips that transmit spikes among them in real-time, while performing some processing. Although these tools reach very high bandwidth at the AER communication level, they require the use of a personal computer to allow the higher level processing of the event information. We propose the use of an embedded platform based on a multi-task operating system to allow both, the AER communication and processing without the requirement of either a laptop or a computer. In this paper, we present and study the performance of an embedded multi-task AER tool, connecting and programming it for processing Address-Event information from a spiking generator.

Keywords Address-Event-Representation, AER tool, embedded AER computation.

3.1 Introduction

Living creatures are able to realize tasks that are not easily done by traditional computation systems. We can receive a huge amount of visual information, distinguish an object in motion, infer its future position and act on our muscles to take it in the order of milliseconds. Neuro-informatics aims to emulate how living beings process data. Efforts are being made in recent years by the research community [1] to develop VLSI chips that perform bio-inspired computation.

N. Martínez Madrid, R.E.D. Seepold (eds.), *Intelligent Technical Systems*, Lecture Notes in Electrical Engineering 38, DOI 978-1-4020-9823-9_3,

Address-Event-Representation, AER, was proposed by the Mead lab in 1991 for communicating between neuromorphic chips with spikes [2]. There is a growing community of AER protocol users for bioinspired applications in vision, audition systems and robot control, as demonstrated by the success in the last years of the AER group at the Neuromorphic Engineering Workshop series [3]. The goal of this community is to build large multi-chip and multi-layer hierarchically structured systems capable of performing massively-parallel data-driven processing in real-time [4]. A deeper presentation of AER will take place in Section 3.2. These complex systems require interfaces to interconnect them and to connect them to PCs for debugging and/or high level processing. There is a set of AER tools based on reconfigurable hardware that can be connected to a computer. They achieve these purposes with a very high AER bandwidth but with the need of a PC for the higher level processing. A new philosophy was born at the last Workshop on Neuromorphic Engineering (Telluride, 2006) to improve this, which is based in the use of an embedded GNU/Linux system running over an embedded powerful microprocessor with network connectivity. This will let neuromorphic engineers to use AER standalone platforms for high level event processing when developing or building AER systems, to use it as a first test platform for hardware implementation of new algorithms and to implement complex algorithms of neuroinspired models which are not always easily portable to pure hardware solution, as learning algorithms, development of connectivity, etc.

We present in this chapter a microprocessor based solution, where the AER bus is connected directly to it by using its general purpose I/O ports, GPIO, as a first approach and in order to study the advisability of its use within AER based systems. We will solve the image reconstruction and edge extraction from event streams problems for this purpose, which requires a high AER bandwidth when no preprocessing is done and will let evaluate the performance of the embedded system. Also, we have compared the proposed solution with other hardware solutions and other multi-task approaches.

3.2 Address-Event-Representation

Figure 3.1 shows the principle behind the AER. Each time a cell on a sender device generates a spike, it communicates with the array periphery. A digital word representing a code or address for that cell is placed then on the external inter-chip digital bus, the AER bus. This word is called event. Additional handshaking lines, Acknowledge and Request, are used for completing the asynchronous communication. In the receiver chip, the spikes or events are guided to the cells whose code or address appeared on the bus. In this way, cells with the same address in the emitter and receiver chips are virtually connected by streams of spikes. These spikes can be used to communicate analog information using a rate code, by relating the analog information to the time between two spikes that correspond to the same neuron, although this is not a requirement. More active cells access the bus more frequently than those that are less active. The use of arbitration circuits usually ensures that cells do not access the bus simultaneously. These AER circuits are generally built using self-timed asynchronous logic [5].

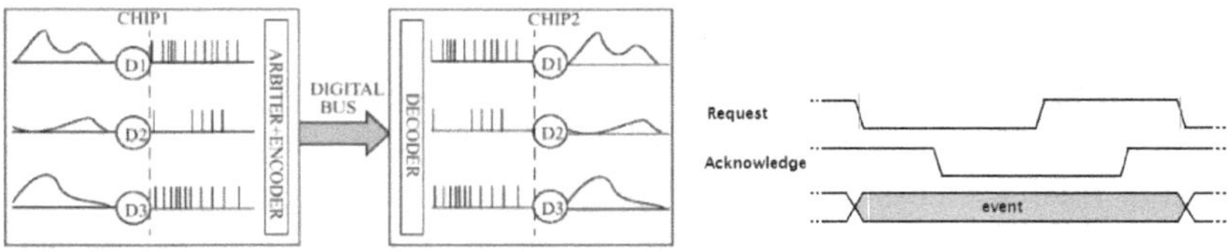

Fig. 3.1 Rate-coded AER inter-chip communication *scheme*

In general, AER is useful for multistage processing systems, in which as events are generated at the front end they travel and are processed down the whole chain (without waiting to finish processing each frame). Also, in multistage systems, information is reduced after each stage, thus reducing the event traffic. A design of a neuromorphic vision system totally based on AER has taken place under the European IST project CAVIAR, "*Convolution Address-Event-Representation (AER) Vision Architecture for Real-Time*" (IST- 2001-34124) [1]. This chain is composed by a 64×64 retina that spikes with temporal and contrast changes [9], two convolution chips to detect a ball at different distances from the retina [6], an object chip to filter the convolution activity [7] and a learning stage composed by two chips: delay line and learning [8]. The maximum throughput rate takes place at the output of the silicon retina. Although it is able to emit 4Mevents/s, real applications, such as someone walking along a corridor or even the beat of an insect wing, vary from 8 to 150 Kevents/s [9], respectively. These values will be used further for comparing the results with real applications (cf. Table 3.1).

Table 3.1 Event Rate for some previous AER-tools. The communication to or from the PC is done by the PCI bus or the USB protocol. They achieve a very high AER bandwidth but with the need of a PC for the higher level processing

AER-tool name	Event rate	AER-tool name	Event rate
Rome PCI-AER	1 Mevents/s	CAVIAR PCI-AER	8 Mevents/s
USB-AER	25 Mevents/s	USB2AER	5 Mevents/s
mini USB-AER	300 Kevents/s		

The research community is also working on applying these systems to different actuators [10–13], from translating AER information into actuator control information (e.g. PWM and PFM) to developing hybrid systems, bioinspired sensors for acquiring preprocessed data and classical computation for decision and control. This hybrid scheme is successfully being applied to other fields, such as sensor networks [14,15].

3.3 Spike Processing Over Multi-Task

Generally, buffers of event streams are prepared on the PC [16], and sent via these AER-tools to the AER bus or an obtained event stream is sent to a PC and a high level processing is done then, such as learning algorithms for the VLSI neuronal network, development of connectivity, models of orientation selectivity, which are not always easily portable to pure hardware solutions [17,18]. Let us present two significant approaches/examples of multi-task spike processing and highlight aspects to bear in mind when translating to an embedded platform.

The first one is interesting because it covers PCI connection and high level spike processing over a GNU/Linux operating system. A hardware/software framework for real-time spiking systems was proposed in [17]. Rome PCI-AER [3] is connected to a Pentium IV at 2.4 GHz running a GNU/Linux 2.4.26 desktop distribution. The software architecture consists on kernel code, module, for controlling this AER hardware interface, an UDP server of event buffers, called monitor, and one or more clients for high level spike processing, called agents. The monitor is implemented in C++ and agents could be implemented in other languages, such as Matlab. It presents a maximum event rate of 310 Kevents/s without high level spike processing.

This second approach covers USB connection to the PC and actuator control at a hybrid processing system under Microsoft Windows XP operating system. Delbruck [13] presents a hybrid system for fast motor control. A single-axis arm acts as goalkeeper and is able to block 80–90% of balls that are shot with >150 ms time to impact. A silicon retina, bioinspired, acts as visual sensor [19]. Spikes are collected by a USB2AER board [20] and sent to a PC, a 2.1 GHz Pentium M laptop, over USB. Data is processed, procedural computation, by a Java application over Microsoft Windows XP operating system and the result is sent over USB full-speed to a motor control board based on C8051F320 MCU, which acts on the single-axis arm. There are three threads: high priority one for reading events from USB, highest priority one for writing motor control decision information to the USB motor control board and one for visualizing the scene and GUI. Each ball generates 30 Kevents/s as mean value, which means an USB Bulk transfer of 128 events every 4 ms.

On both approaches, powerful PCs are used, so it is not possible to implement their systems as embedded standalone ones. Also, complex and efficient processor architectures are used. Embedded systems do not have so resources. On the first approach, there is a data and time overhead from the UDP/IP protocol architecture, even if loop-back network interface is used. It would be desirable to avoid it due to the resource limitations of embedded systems. The second approach runs on JVM which introduces instruction overhead and reduces the performance. Although only some part of the application is running purely on it, byte codes have to interpreted and corresponding processor instructions have to be executed, garbage collector may execute periodically, etc. Once more, the use of JVM should be avoided to get a better performance on an embedded system.

3.4 Embedding Multi-Task Spike Processing

We propose the use of Intel XScale PXA255 processor running an embedded GNU/Linux 2.6 system using uClibc and double-buffering signaled exchange scheme for receiving and processing AER information.

3.4.1 Hardware Architecture

The Intel XScale core [21] is an ARM V5TE compliant microprocessor and provides the ARM V5E DSP extensions, although it does not provide hardware support for floating point instructions. It is a 7-stage integer/8-stage memory super-pipelined core. The core presents a Multiply/Accumulate unit, MAC unit, that supports early termination of multiplies/accumulates in two cycles and can sustain a throughput of a MAC operation every cycle. Also, it offers 32 KB of data cache, 32 KB of instructions cache and an MMU. The ability to continue instruction execution even while the data cache is retrieving data from external memory, a write buffer, write-back caching, various data cache allocation policies which can be configured different for each application and cache locking improve the efficiency of the memory bus external to the core. In addition, a Branch Target Buffer is present, that holds 128 entries with a miss predicted branch latency penalty of 4 core cycles and 0 when predicted correctly. The processor has 84 GPIOs that can be programmed to work as function units to manage serial ports, I2C, PWM, LCD, USB client 1.1, etc. In addition, platform will need RAM, Flash memory and network connectivity.

3.4.2 System Software Architecture

uClibc is a lightweight and widely used library for developing embedded Linux systems, which supports shared libraries and threading. This lets the application's binaries to be lighter and allows running on tiny hardware systems. GNU/Linux is a multi-task general purpose operating system, so it is designed for obtaining a good mean performance. So, it needs to be adapted for AER computing.

AER was developed for multiplexing in time the spike response of a set of neuro-inspired VLSI cells. Neuro-inspired cells are not synchronized. They send a spike or event when they need to send it and the AER periphery is responsible to send it into AER format with the minimum possible delay, and therefore, the AER scheme is asynchronous. Although handshaking lines guarantee the delivery of an event, if the receptor stalled the AER communication it can cause to process information from the past and not the up-to-date one. Therefore, it is desirable the shortest response time. A more fine-grained resolution system can be achieved by rising the frequency value of the timer interrupts, which not only implies a shortest process response time but a quicker turnover of scheduler's processes queue. On the other hand, an extra instruction overhead has to be paid due to a higher number of timer interrupts. This implies context switches from process to interrupt handler and from this last to the first, the handler execution, and possible cache and TLB pollution, which may result in an impoverishment of the system performance. This value is set before the Linux kernel compilation process.

The scheduling policy determines how the processes will be executed in a multi-task operating system. The Linux kernel 2.6 version presents several ones. The kernel offers system calls to let the processes to choose the scheduling policy that will rule their execution. A dynamic priority based on execution time scheduling policy, a real-time fixed priority FIFO one and a real-time fixed priority round robin one are offered by the kernel. The first one is the common policy on UNIX systems. Basically, a base priority is initially assigned to the process based on the frequency value of the timer interrupts. Its new priority is calculated by the scheduler when this last is executed using the execution time associated to the process. This priority will determine when the process will be executed again. The other two scheduling policies differ from each other in how processes with the same priority are reorganized to take the microprocessor again, using a FIFO criterion or a round robin one, respectively. A process whose execution is managed by one of these two policies is, obviously, not influenced by the first of all. Even more, preference will be given, of course, to a process in these scheduling situations than the managed by the first policy ones. The real-time scheduling policies try to ensure a short response time for a ruled by them running process, which is desirable when development an AER device. Also, no lower-priority processes should block its execution but this situation actually happens. The kernel code is not always assumed to be pre-emptive because it has to be compiled with this option and it is only supported in 2.6 versions. So a system call from a lower-priority process may block the execution of higher-priority one until it has finished. Therefore, the support for real-time applications is weak although the processes response time is improved referred to the common scheduling policy. Every process in a Linux system is normally ruled by the first one. Therefore, a process running continuously cannot be set to be ruled by one of the offered real-time policies without making the whole rest of the system unresponsive. If other processes e.g., network,... are needed, a combination of the scheduling policies at runtime based on the application state, receiving events or waiting for them, could increase the performance of the system with no degradation on the multi-task environment response.

3.4.3 User Software Architecture

High level spike processing is not applied individually to one event but a set of them. So, we propose a double-buffering scheme for the AER communication and this high level event processing on this system, splitting up both into two concurrent tasks, trying to make the most of the time between event arrivals for spike processing. Also, this separation makes the development of this kind of applications easier. Only special spike processing has to be developed due to the AER communication is obviously always the same.

Special care must be taken when defining the buffer size in events unit. If it is too big, it will represent data for a big period of time and the information that is being processed may differ a lot from the current output of the emitter, which may cause to take a not valid decision for the current state of the whole AER application/system. Time continuity must be guaranteed at the virtual parallelism level. In addition, event arrival timestamps are collected when each is received. Linux provides 64 bit integer variable, called jiffies that are incremented on each timer interrupt. Inter-Spike-Interval, ISI, may be two or three orders lower, so we will use a processor 32 bit timestamp counter [21]. It is incremented on each core cycle, so ns resolution is provided. It can only be accessed by privileged instructions, so a kernel privileges is need. A Linux module will be responsible for obtaining the event timestamp. Although Linux 2.6 system calls are not always preemptive, as mentioned before, obtaining the value of the time-stamp counter is an atomic processor instruction and it is guaranteed the ending of its execution (cf. Fig. 3.2).

As the event arrival is asynchronous, the event buffer filling is also asynchronous. We propose the use of signals, which are asynchronous too, for notifying the double-buffering buffer exchange. When a process receives a signal, it processes the signal immediately, without finishing the current function or even the current line of code. The operating system stops its execution and assigns the processor to the signal handler that has been registered for that signal. A signal handler should perform the minimum work necessary to respond the signal and return control to the main program then. So, we suggest a buffer references exchange to the appropriate buffer depending on the received signal as the signal handler. In this way, up-to-date data is quickly able to be processed. Therefore, both approaches described in Section 3.3 present an extra overhead with regard to our proposal. In the first one, context switches for the kernel code of the module for the AER hardware interface, the monitor and agents, demanding many resources when implemented on high level languages such as Matlab, and the UDP/IP protocol architecture overhead should be considered. In the second one, USB Bulk transfers of 128 events are used. Microsoft Windows XP USB controller checks USB interrupts each 1ms. Interrupt service and transferring data from kernel to user space should be also considered. Events can be received each 1 μs [19], so the buffer will be ready at the USB2AER in 128μs and will be ready to be processed after more than 1ms. During this time more than 872 events can be received at the USB2AER board.

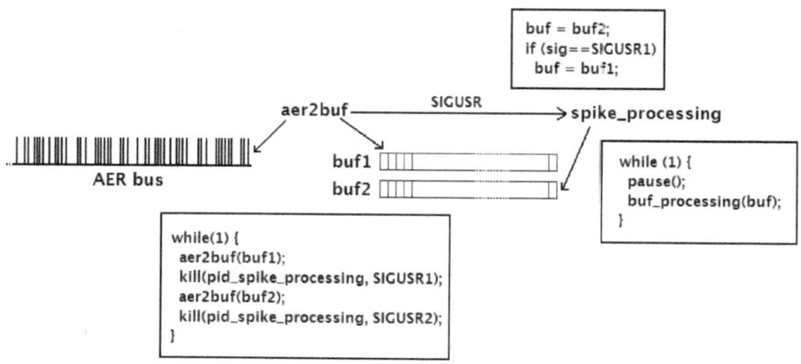

Fig. 3.2 Software architecture for high level spiking processing over a multi-task system when receiving events from the AER bus

High level spike processing will be applied to a set of events, so it will be done when a buffer has been filled. There are three possibilities based on the signal communication and the task latencies. If filling the buffer, either at AER communication level or computing them, lasts more than consuming it, every event will be treated. If not, the consuming task will work with the last updated events and, together with the rate-coded AER's feature of "losing some events does not necessary mean losing information", it does not implies to be always an undesirable situation. Finally, a returned signal from the task that consumes the buffer could be added to the scheme, as a "ready signal", ensuring the processing or the reception-emission of every event, independently of the latencies of both tasks.

The processor offers a mechanism to detect any level change at any of its GPIO ports, generating hardware interrupt when it occurs with a minimum pulse width duration to guarantee this detection is 1 μs [21]. It is necessary to detect the two Request signal levels to implement the AER hand-shake protocol. In addition, over 0.17 μs are needed to set a bit on a GPIO in this processor. Two sets have to be done for generating the AER Acknowledge signal. Therefore, the minimum time between events would be, at least, 2.34 μs. It should be greater considering the time penalty due to the interrupts handlers execution, context changes. . . which implies a event rate fewer than 427 Kevents/s only for the AER communication task. AER communication is asynchronous, so either the number of consecutive events or the time between two of them cannot be supposed. Free spikes or bursts of them can appear in the bus. Although hardware interrupts release the processor for computation tasks until data is ready at I/O, if spikes are presented as bursts of events the event rate will be reduced. Also, if there is no event traffic at the AER bus for a period of time enough long, there is no high level spike processing to do and so, there is no need to release the processor. Therefore this option may be ruled out, and polled I/O may be used.

From a computational point of view, both, filling a buffer from the AER bus or sending to it, makes the AER communication to be a worst-case linear time algorithm. The proposed double buffering buffer exchange is a worst-case constant time algorithm. Therefore, this software architecture presents worst case linear time complexity, whose worsening may only take place at the high level spike processing task.

3.5 Image Reconstruction and Edge Detection

We will use two tasks for testing the system: (1) reconstructing a frame from an event stream and (2) edge detection. This last is done by applying convolutions with a common 3×3 kernel matrix for this purpose, so is a worst-case polynomial time algorithm, of complexity $O(n^3)$. This will let the system performance to be tested with a more complex algorithm.

In artificial vision systems based in AER, it is widely used the rate-coded AER. In this scheme, each cell corresponds to a pixel and its activity is transformed into pixel event frequency. This scheme may be inefficient for conventional image transmission: Monochrome VGA resolution (480×640 pixel frames, at 25 frames per second, with 8 bits per pixel) yields a peak rate of (480×640 pixels/frame) × (256 spikes/pixel) × (25 frames/s) × (19 bit/spike) = 37 Gbit/s. So preprocessed images are usually transmitted instead of raw images, such as edges or contrast [22], and therefore, previous full VGA peak rate is reduced in two or three orders of magnitude. Another one or two orders can be added in this reduction due to the use of image resolutions between 64×64 and 128×128 pixels at the most. Even then, image reconstruction from rate-coded AER presents a very high demanding through-output. That is why we choose it for testing the performance of the system, although this is not a multimedia protocol. It is a visual information processing scheme. Going from asynchronous AER to synchronous frame based representation video is more or less straightforward. If T_{frame} is the duration of a single frame, a 2-D video frame memory is reset at every time $t=n \times T_{frame}$, where $n \in [0, \infty)$, called the integration time. Then, for each event address, the memory position for this address (x, y), is incremented by 1. Finally, the content of the 2-D memory is transferred to the computer screen and reset again at $t=(n+1) \times T_{frame}$. This is more or less how state-of-the-art AER hardware engineers visualize their AER vision systems outputs on computers [23].

We have developed a processes and a threads implementation for the tasks mentioned before. We use IPC Shared Memory method in the first one and global variables in the second one for the shared data, which makes both implementations equivalent from the access to memory point of view. For both implementations, we propose an AER-communication driven execution policy with no "ready" like signal, which will decide the execution rate. So, events will be continuously collected and put into a buffer, "aer2buf". When this buffer is full, a signal will be sent to the high level processing task and new received events will be put into the other buffer. The spike processing task, "buf2img", will be generating the frame into memory from a buffer or waiting to receive a signal, so it will only consume processor execution when it is needed. Therefore, it is also a worst-case linear time algorithm which let to continuously generate the frame or wait until a buffer is ready for its treatment. For the edge detection implementation, kernel convolution will be applied once the frame is constructed. Finally, buffer size has been set to 200 events.

3.6 Results

We will use a small factor size (80×20×5.9 mm) Intel PXA255 400 MHz board running GNU/Linux 2.6 using uClibc, Gumstix Connex-400 motherboard. It also offers 64 MB of RAM and 16 MB of Flash Memory. Gumstix Wifistix board will provide IEEE 802.11 connectivity. An USBAER [24] board will play the role of the AER emitter. It will be responsible to transform a binary representation of a frame into the corresponding events and to send them using the exhaustive synthetic AER generation method [25]. These will be sent to the platform via the AER bus, whose pins will be directly connected to the processor's GPIO ports. The frame is downloaded to the USB-AER from the PC, no preprocessing is done, such as referred in Section 1, and it will continuously be sending the same frame translated into event streams. This board is able to achieve an event rate up to 25 Mevents/s. Having this event rate will let us to evaluate the performance of the embedded computer, which should be the bottle-neck. An oscilloscope probe will be clipped to the Request signal pin and it will be used to measure the event rate, due to each cycle at this signal implies an event communication. The usual mechanisms to compute the execution time of a task and its duration, either provided by the hardware or the operating system, would interfere on the obtained value by incrementing it. So the need of including this kind of instructions is avoided by using the oscilloscope. The event rate will be the frequency of the Request signal, which will be calculated by it. The time during the process is ready to run and waiting to take the processor for its execution is also considered in this value. This makes it a real measure of the mean even rate for AER communication and spike processing. Finally, another process will be used for debugging purposes, independently of the double buffering implementation. This process will be waiting to receive a signal that will be periodically sent by the operating system. Then, it will wake up and put the frame in memory into a BMP file. This last can be viewed by connecting to the HTTP server on the platform. Also, these processes will be used to test the implementations under situations with other ones running.

We have executed both implementations, processes and threads, for image reconstruction and edge detection and studied their evolution over the time. The threads implementation achieves an event rate, ER, of 770 Kevents/s, while the processes one reaches 540 Kevents/s. These values are reached even if there are other processes running on the system and are mainly maintained over the time. Both implementations present a momentary reduction of the ER. When no other process is running, these worst event rates, WER, are 620 Kevents/s for the threads implementation and 450Kevents/s for the other one. These oscillating values define event rate intervals that are relatively small but WER evolves sometimes to a harsh value of 259 Kevents/s and 200 Kevents/s for each implementation, respectively, when there are other processes running on the system. Although these last WER values appear momentarily, they suppose a main degradation of the spike processing performance.

We have increased the frequency of the timer interrupts from 100 Hz, the default one for the ARM architecture, to 1000 Hz. At this one, the event rate is not affected by the fact of other processes running on the system. The ER is 775 Kevents/s and WER is 660 Kevents/s for the threads implementation and 500 Kevents/s and 430 Kevents/s, respectively, for the other one. So, it has been achieved that the influence of other processes on the event rate is transparent for a frequency value of the timer interrupts of 1000 Hz.

We have set our threads implementation to be ruled by the real-time fixed priority round robin scheduling policy, achieving an event rate of 840 KEvents/s continuously maintained over the time. Therefore, the time between two consecutive events is 1.19 μs. This value is near the best one, but as we have explained before, and so expected, the system was unresponsive for other tasks.

We have also measured the exact time between events for the system using the oscilloscope, which is 1.16 μs. Therefore, the system presents an event rate of 862 Kevents/s without either the spike processing task or other processes running on the system. The threads implementation presents 770 KEvents/s, which implies that it performs the event acquisition and the event treatment with a mean time between events of 1.29 μs, approximately. Therefore, it offers a multi-task environment useful for other simultaneous tasks with an 11% deviation from the maximum that can be achieved with the system. Under the real-time round robin scheduling policy, the mean time between events is 1.19 μs, so spike processing implies a 2.5% from the maximum. In Section 3.2, a neuromorphic vision system totally based on AER has been presented. The maximum throughput rate takes place at the output of the silicon retina and varies from 8 to 150 Kevents/s for real applications [9]. The higher demanding value, 150 Kevents/s, implies a mean time between events of 6.66 μs. The time of the reception of an event of our system is 1.16 μs. So, there is a mean time of 5.5 μs for any kind of high level spike processing, which means up to 2200 instructions on this 32-bit processor at 400 MHz.

3.7 Conclusion

We have presented an embedded multi-task architecture that allows spike processing at 840 KEvents/s. Its reduced size, the possibility to have other services running simultaneously (network communication) and its PWM outputs for motor control makes it very suitable for standalone hybrid AER systems. The proposed software architecture exploits the platform performance and lets neuromorphic designers to quickly and easily develop new applications.

Acknowledgments We would like to thank the NSF sponsored Telluride Neuromorphic Engineering Workshop, where this idea was born in a discussion group participated by Daniel Fasnacht, Giacomo Indiveri, Alejandro Linares-Barranco and Francisco Gómez-Rodríguez. This work was supported by Spanish grant TEC2006-11730-C03-02 (SAMANTA 2).

References

1. R. Serrano-Gotarredona, M. Oster, P. Lichtsteiner, et al., "AER building blocks for multi-layer multi-chip neuromorphic vision systems". Neural Information Processing Systems Conference, 2005.
2. M. Sivilotti,, "Wiring considerations in analog VLSI systems, with application to field programmable networks". PhD thesis, California Institute of Technology Pasadena, CA, USA, 1991.
3. A. Cohen, R. Douglas, C. Koch, et al., Report to the National Science Foundation: Workshop on Neuromorphic Engineering, 2001.
4. M. Mahowald, "VLSI analogs of neuronal visual processing: a synthesis of form and function". PhD thesis, California Institute of Technology, 1992.
5. K. Boahen, "Communicating neuronal ensembles between neuromorphic chips", Neuromorphic Systems Engineering, 1998.
6. R. Serrano-Gotarredona, T. Serrano-Gotarredona, A. Acosta-Jimenez, and B. Linares-Barranco, "An arbitrary kernel convolution AER-transceiver chip for real-time image filtering. Circuits and Systems", ISCAS 2006. Proceedings. 2006.
7. M. Oster, and Liu, "A winner-take-all spiking network with spiking inputs", IEEE Conference Electronics, Circuits and Systems, pp. 203–206, 2004.
8. H. Riis, and P. Hafliger, "Spike based learning with weak multi-level static memory. Circuits and Systems", 2004. ISCAS'04, 2004.
9. P. Lichtsteiner, and T. Delbruck, "64×64 Event-driven logarithmic temporal derivative silicon retina". IEEE Workshop on Charge-Coupled Devices and Advanced Image Sensors, pp. 157–160, 2005.
10. A. Linares-Barranco, F. Gómez-Rodríguez, A. Jiménez-Fernández, et al., "Using FPGA for visuo-motor control with a silicon retina and a humanoid robot", IEEE International Symposium on Circuits and Systems, pp. 1192–1195, 2007.
11. A. Linares-Barranco, A. Jimenez-Fernandez, R. Paz-Vicente, et al., "An AER-based actuator interface for controlling an anthropomorphic robotic hand", LNCS, 4528:479, 2007.
12. A. Jimenez-Fernandez, R. Paz-Vicente, M. Rivas, et al., "AER-based robotic closed-loop control system", IEEE International Symposium on Circuits and Systems, pp. 1044–1047, 2008.

13. T. Delbruck, and P. Lichtsteiner, "Fast sensory motor control based on event-based hybrid neuromorphic-procedural system", IEEE International Symposium on Circuits and Systems, pp. 845–848, 2007.
14. T. Teixeira, E. Culurciello, J. Park, et al., "Address-event imagers for sensor networks: evaluation and modeling", International conference on Information processing in sensor networks, pp. 458–466, 2006.
15. D. Bauer, A. Belbachir, N. Donath, et al., "Embedded vehicle speed estimation system using an asynchronous temporal contrast vision sensor", EURASIP Journal on Embedded Systems, 2007(1):34–34.
16. A. Linares-Barranco, G. Jimenez-Moreno, B. Linares-Barranco, and A. Civit-Balcells, "On algorithmic rate-coded AER generation", IEEE Transactions on Networks, 17(3):771–788, 2006.
17. M. Oster, A. Whatley, et al., "A hardware/software framework for real-time spiking systems", Int. Conf. on Artificial Neural Networks, 3696:161–166, 2005.
18. E. Chicca, A.M. Whatley, P. Lichtsteiner, et al., "A multichip pulse-based neuromorphic infrastructure and its application to a model of orientation selectivity", IEEE Transactions on Circuits and Systems, 54(5):981–993, 2007.
19. P. Lichtsteiner, C. Posch, T. Delbruck, "A 128 X 128 120db 30mw asynchronous vision sensor that responds to relative intensity change. Solid-State Circuits", IEEE International Conference Digest of Technical Papers, pp. 2060–2069, 2006.
20. R. Berner, T. Delbruck, A. Civit-Balcells, and A. Linares-Barranco, "A 5 Meps $100 USB 2.0 address-event monitor-sequencer interface", IEEE International Symposium on Circuits and Systems, 2007.
21. Intel-Press, Intel PXA255 Processor Developer's Manual, volume 278693-002. Intel-Press, 2004.
22. K. Boahen, and A. Andreou, "A contrast sensitive silicon retina with reciprocal synapses", Advances in Neural Information Processing Systems, 4:764–772, 1992.
23. E. Culurciello, R. Etienne-Cummings and K. Boahen, "A biomorphic digital image sensor". IEEE Journal of Solid-State Circuits, 38(2):281–294, 2003.
24. R. Paz, F. Gomez-Rodriguez, M. Rodriguez, et al., "Test infrastructure for address-event-representation communications", Work-Conference on Artificial Neural Networks (IWANN'2005). LNCS, pp. 518–526, 2005.
25. F. Gomez-Rodriguez, R. Paz, L. Miro, et al., "Two hardware implementations of the exhaustive synthetic AER generation method". Computational Intelligence and Bioinspired Systems. LNCS, 3512:534–540, 2005.

Chapter 4

End to End UPnP AudioVisual Service Provisioning and Management

Javier Martínez, Natividad Martínez Madrid and Ralf Seepold
Universidad Carlos III de Madrid, 28911 Leganés (Madrid), Spain,
javier.martínez@uc3m.es, natividad.martínez@uc3m.es, ralf.seepold@uc3m.es

Abstract The Universal Plug and Play (UPnP) technology is widespread among many multimedia consumer electronic devices and applications, but its usage is typically restricted to home (local) networks. The objective of this work is on one hand to extend the use of UPnP beyond home networks with the help of the Session Initiation Protocol (SIP) technology and on the other hand the remote management of the Media Servers through OSGi technology. The proposed architecture allows the distribution of devices and services, keeping the same application functionality and usability. Thus, it is possible to transparently discover and invoke services in devices from two different home networks, creating end to end connections. The development has been implemented over IPv6; explaining all required changes in firmware and software, while keeping compatibility with IPv4 so that future changes are transparent for the user.

Keywords Middleware, Service provisioning, Remote Management, OSGI, UPnP.

4.1 Introduction

Protocols like SIP [1] or UPnP [2] allow detecting and accessing different elements in a network independently of their localization. UPnP Media Architecture [3] and SIP permit to access and exchange multimedia content like sound or video.

However these protocols solve the localization and access problem in completely different ways. In SIP a unique identifier is needed to find the destination and is centered in any kind of communication sessions. On the other hand UPnP (Audio Video part) is based in the use of broadcast messages to discover the different services available in the local network and

N. Martínez Madrid, R.E.D. Seepold (eds.), *Intelligent Technical Systems*, Lecture Notes in Electrical Engineering 38, DOI 978-1-4020-9823-9_4,

in transmission of multimedia content using mainly unidirectional (streaming) communications.

Interoperability can be established between both protocols to solve the different access problems of each one. The main idea is to encapsulate an UPnP service so that it can be offered to the system as an identifiable service and easily localizable using SIP.

This way the UPnP devices can be delocalized. The impossibility of using this protocol to provide an end to end solution can be solved using SIP.

The idea of working with protocols using IPv6 [4] comes from two different needs. The first one is the imminent migration to IPv6 where there is no clear support at hardware and software level. This way the development is already adapted to use this protocol. The other is the growing use of personal devices at home where the use of IPv4 creates problems like the use of NAT where IPv6 does not.

Apart of this, but with total relation, is to get the modularity of all parts. In this case, it is possible to manage remotely all components like Media Servers, SIP Modules, etc. And this is a great advantage to the user because it supports the idea of delocalization of SIP. This is one of the reasons because all modules are programmed under OSGi platform.

The next section introduces the state of the art with a brief description of the used technologies and some studies in this area. Then it will be described the environment in which this work has been developed and tested. Then the whole end to end process will be described. This paper finalizes with conclusions and a description of possible future works.

4.2 State of Art

Several studies like Vallejo [5] foresee the status of IPv6 and the need for its implementation.

There are some works that try to provide end to end service using UPnP. Nikolaidis [6] tries to combine UPnP with other technology like CWMP, but in this case the work is based in remote management made by the service provider but does not deals with end to end between different users.

Another end to end example is the one proposed by Ditze [7]. Although in this case an extension to the Quality-of-Service (QoS) part of the UPnP standard is proposed to allow end to end connections. But no solution is given to provide services from a residential network to another.

Vilei [8] uses both UPnP and SIP to create a videoconference system, but the objective if far from providing services from a home to another. On the same line Yukikazu [9] uses SIP again with some UPnP functionalities to discover devices, but more as a way to implement the UPnP protocol using SIP.

Without a doubt mobility and delocalization are the main advantages of SIP as shown by Brown [10]. Although he uses devices with the OSGi Framwork [11] and uses SIP for delocalization.

In the other hand, the integration of networks for discovery of devices and services in residential networks with the support of OSGi platforms have been published in 2002 [12]. Currently, there is an implementation independent UPnP distribution for OSGi devices available (Domoware [13]). Bundles of the OSCAR framework for OSGi are integrated in this repository. Also Kang [14] published a work for integration of the audio-visual architecture UPnP AV in an OSGi platform.

4.3 Technologies Overview

Therefore, four basic technologies are used to extend the current situation and to support the integration of services: Firstly, the universal plug and play (UPnP) protocol standard is presented briefly. In this paper we focus in the share of multimedia content and in this way there is a specific standard of UPnP focus in multimedia the UPnP AV Architecture. Here, a brief introduction is given. Secondly, the platform is developed in IPv6, so is needed to explain the current state about IPv6 and how is the current implementation of this protocol, and finally we introduce some concepts of SIP and OSGi.

4.3.1 UPnP

One of the objectives is to access multimedia content in LANs. The UPnP AV architecture provides a control mechanism to support sharing multimedia content in different types of devices, formats of content or transfer protocols.

The UPnP technology, developed by the UPnP forum, allows devices to form communities and to share services without an increase in complexity of configuration or demanding for human intervention. In an UPnP network, a device automatically discovers other devices.

The UPnP AV architecture (version 2) defines interactions between an UPnP control point and multimedia devices. In general, all the interactions between the devices involved are controlled through control points. The multimedia communication setup is controlled through control point, but the transfer of content is directly transmitted from devices servers to the reproducers, as it is shown in Fig. 4.1.

The UPnP MediaServer and the UPnP MediaRender [15] are two UPnP standard devices. The control point, is not standardized by UPnP since it itself does not offer any interface.

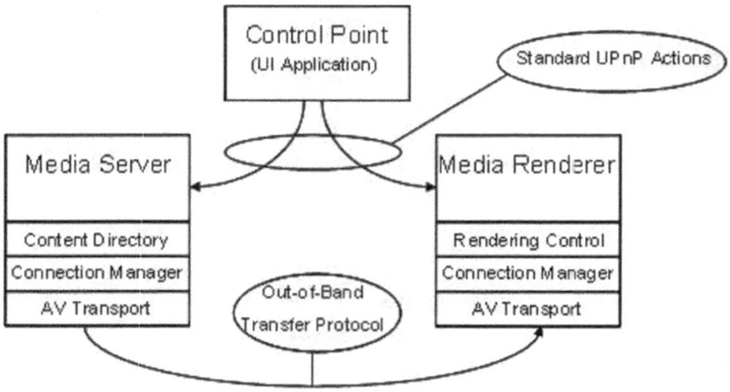

Fig. 4.1 Interaction model of UPnP AV devices

4.3.2 OSGi

The OSGi specification provides a platform for the development of applications and services widely supported by companies of different sectors. The core of the specification defines a framework that provides an environment for downloading and execution of software that offers services. Based on Java, this solution guarantees the independence of the hardware platform and operating system, with few software requirements.

OSGi provides an execution environment that facilitates the interoperability of devices. It offers a mechanism for the definition of services in a standard and modular form, so that they can be used like component blocks for other services. The OSGi Framework allows the dynamic installation of new services, as well as starting, shutdown, remove and update.

These services are independent of their hardware platform and applications or the underlying operating system since they are executed on a Java virtual machine. The packages with the code are modules called bundles. A service can consist of only one bundle, or can be compound of any number of them. Bundles communicate with each other one and offer their services to others bundles, creating therefore a chain of dependencies. Fig. 4.2 shows this architecture.

4.3.3 IPv6

IPv4 protocol is getting older since several years ago. This represents a problem when new services are to be implemented in the internet, like end to end connections. The lack of IPv4 addresses for all the devices connected to the internet and the need to use NAT (Network Address Translation) with local (private) addresses breaks these end to end connections.

The possibility to offer an IPv6 address for every device connected to the internet, including devices in a home network, makes IPv6 one of the better and most suitable solutions for end to end connections implementations.

IPv6 provides three kinds of addresses. Link-local addresses are to be used only at link level. But site-local and global addresses can be routed and so can be used normally by applications. Site-local addresses are designed to be used only at a site level, substituting the use of private IPv4 addresses. Global IPv6 addresses are to be globally routed in internet. Even device in home networks can make use of a global address, allowing end to end connections.

Multicast is part of the base specifications in IPv6, unlike IPv4, where it was introduced later. The use of broadcast and link-local addresses is the default way to work for UPnP in local networks with IPv6. The solution proposed in this work is not based on multicast or broadcast through internet but in then use SIP to connect two different Control Points in different networks using a global unicast address. The IPv6 implementation on internet and home networks is on the way so adapting application protocols and networks is mandatory to support future services.

4.3.4 SIP

SIP is a signaling protocol developed to initialize, modify and terminate a user interactive sessions characterized by the intervention of multimedia elements like video, voice, instant messaging, on-line games and virtual reality.

This protocol offers personal and terminal mobility over WANs, so that any device that incorporates a sip agent can be located and connected wherever it is. In addition, this protocol solves problems related to the devices handover and multihoming.

In the other hand, SIP protocol will offer extended features related to authentication, encryption and quality of service control that supplies the capabilities to establish an external communication between two different sub domains.

Other SIP parallel technologies at the mobility management point of view are DDNS [16] and MOBIKE [17]. In one hand, DDNS is based on the use of dynamic DNS's that are updated their content. In contrast, this technology doesn't support such as handover as multihoming. In the other hand, MOBIKE (IKEv2 Mobility and Multihoming) is based on the use of the signaling protocol that integrates IPsec, IKEv2. This protocol is based on the Security Associations that are created by the devices IPs. This one provides handover and multihoming support, but this technology does not provide the ability to locate a specific device into different domains. Given that, both technologies were discarded, finally selecting the SIP protocol.

4.3.5 Related Work: Available UPnP AV Developments

There are few implementations available that provide UPnP AV support, and furthermore, these implementations have several incompatibilities. A study has been performed in order to check the available open source tools, since one objective of the work is to reduce costs for the applications connected to the residential gateway and to be able to modify parameters in case it is required. A modification may be required in case of integration.

The following list summarizes some results obtained:

- CyberMediaGate (version 1.2): Is an implementation of a Media Server and is available in Java and in C++, both as open source, and at this moment is the reference implementation for most of the present developments [18]. This Java implementation has been selected for the developments presented in this paper.
- Intel (version 1.0.1768): Provides a Media Server as well a Media Renderer, both of them are open source [19]. The Software includes addition several tools like a sniffer to monitor UPnP communications in the network. Intel also provides a version for Pocket PC (tests have been made in a HP IPAQ 5450).
- Cidero (version 1.5): This is an open source implementation of the UPnP AV Control Point. Cidero uses the code of the CyberMediaGate, the interface is more user-friendly, and it has some additional functions that allow the visualization the available Media Servers and Media Renders. The additional features have been very useful for the demo setup. Although Cidero [20] does not have an own reproducer, it supplies a control for the Media Server and the Render, the reproduction can be controlled from the Control Point. The implementation of Cidero is available for both platforms, Linux and Windows.

- Nokia Media Streamer (version 1.2–12): It is an implementation of Nokia for the Nokia 770 device that provides a Control Point and in a Media Render. This implementation is based on the implementation of S. Konno (Cidero) in C++. The implementation allows to convert the mobile device into a Control Point and it extends its possibilities to be used like Remote Control, for example in case of a Media Server or a Media Render that are available as UPnP devices [21].
- Windows Media Player (version 11): The current version of the player of Windows [22] has the functionality of the Windows Media Connect integrated. The Media Server is serving the multimedia contents within the network UPnP.
- GmediaRender (version 0.0.3): On the Linux platform, there is the only one Media Render available. The installation turned out in a complex task. It does not have graphical interface; the implementation is developed in C. Ongoing development of other projects will add UPnP support. The GmediaRender distribution can be found through Web [23].
- Other new devices: Windows Vista and the Xbox 360 provide Media Server capabilities. Nokia also extends the UPnP AV technology in its new generation *N* of mobile phones.

4.4 Model for Remote Audiovisual System

The model representing the architecture is shown in Fig. 4.2. Both residences are represented. Every residence has its own residential gateway and UPnP devices. The Residential Gateway use to include the AV Control Point and the SIP Agent. This SIP Agent permits to register the Control Point in a SIP Proxy server to exchange SIP messages. With this SIP connection the information about UPnP Devices can be exchanged between Control Points. Once the Control Point has the global IPv6 address of the remote device this device can be accessed directly. UPnP Devices from different residences are the extremes of the end to end connection.

The model implementation has been made using IPv6 from the beginning, keeping compatibility with IPv4. This is due to the need for the residential Gateway to be ready for the Next Generation Networks and be compatible with then in the future.

It represented an extra effort in the development of this work to find the needed tools to work with IPv6 address at any level.

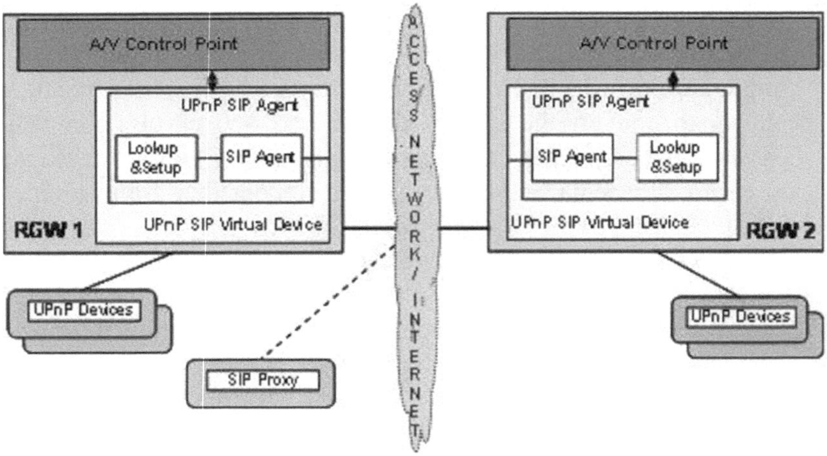

Fig. 4.2 System Architecture

For the implementation of the prototype, the following components had been selected: Control Point of Cidero and the Media Server Cybermediagate, because the developments are open source and both are available for Windows and Linux.

In principle, the Control Point of Cidero and the CyberMediaGate Media Server work like independent applications. This means that, among other things, although initially they can make use of the same UPnP libraries, these libraries are duplicated at the same time of their installation in the machine and in memory at the execution time.

In order to start, an integration of the control point and the server into the OSGi platform has been done: the Control Point developed by Cidero has been transformed in a Control Point Bundle and the Media Server into a Media Server Bundle.

When creating a bundle of an application, the functionality of the application must be encapsulated. This is required due to the requirement to control the bundle by a class that is called BundleActivator. This class includes the methods start and stop to the service when they are invoked. When implementing an application as a bundle, the functionality will be offered like a service, so that this service can be invoked and controlled by others local or remote bundles without using the graphical interface.

The libraries are shared that were previously duplicated in both applications. Therefore, the Control Point and the Media Server use the resources of the residential gateway in a more efficient way.

It has to be noted that there is no Media Renderer implemented for PC, and no one of the implemented supports IPv6. So a *Mini Renderer* has been developed in Java with minimum functionality working with IPv4 and IPv6.

- The following parts have been used to develop this Renderer:
- The Cyberlink UPnP stack modified to work correctly with IPv6.
- Parts of the libraries included in the Cidero Control Point that simplify the implementation of UPnP services like AVTransport, Connection Manager and Rendering Control.
- Java Media Freamework (JMF) [24] which allows reproducing audio and video using different codecs. A Sun codec released later to reproduce specifically MP3 audio format [25] was included too.

4.5 Scenario

In this section it is described the proposed use case and how all the work has been realized. The use case is as follows.

Two different users want to share media content over internet while every user is at a different home. One of the users will share this content using Cyber Garage Media Server so that the other user can see this content in its own residence user a Media Renderer.

Now the whole process will be described to achieve this desired result.

4.5.1 Device Discovery

The first to be done is to lunch both bundles Control Points so that they can detect local devices in their respective networks.

This detection is transparent for the user as it is the way UPnP is designed to work. Every user will see the devices found in its respective Control Point applications. The first user may see how the CyberGarage Media Server bundle is found in the network (see Fig. 4.3).

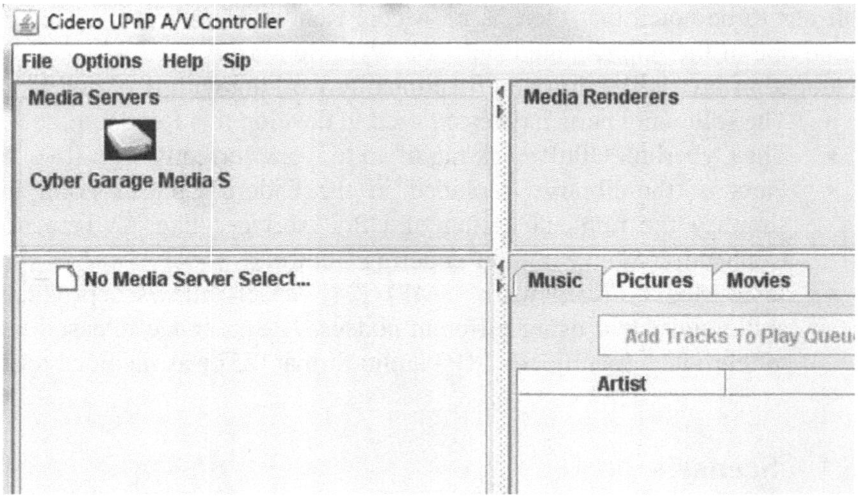

Fig. 4.3 Local detection of UPnP AV devices by the cidero control point at home A

The second user will see how its Control Point detects the renderer but no media content server (see Fig. 4.4).

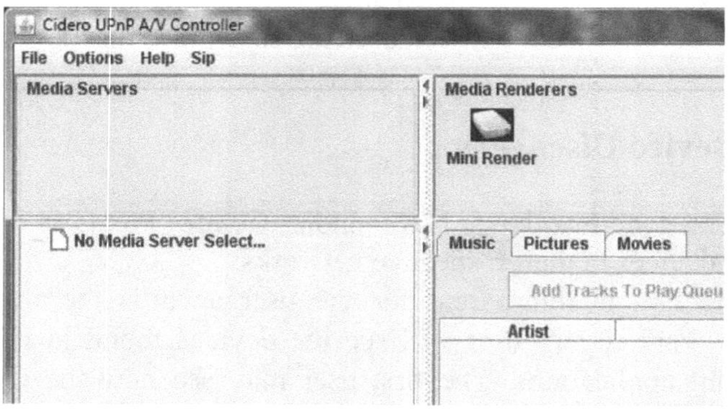

Fig. 4.4 Local detection of UPnP AV devices by the cidero control point at home B

4.5.2 Virtual Devices Transfer

The problem with the actual state of UPnP is how to access to devices in other residences.

To see services from other residences localy, like media servers, to get those devices to be registered in the local Control Point. This way all devices in both residences could be controlled from only one residence.

In case of following only the UPnP standard, the following problems would be found:

1. UPnP devices announcements are made using broadcast but these announcements will not go across the residence router or gateway: provider will not permit broadcast packets to flood internet. The solution to this problem is to send the public addresses of the devices to the other control points using SIP.
2. In case of using IPv4 this wouldn't solve the problem because the used addresses are private and useless in the other home network. The solution to this problem could be using NAT, assigning a public IP and port to a local private address in the residential gateway and then sending the public IP and port to the other control point using the SIP solution. With this solution the router configuration must be modified constantly, representing an expensive (in resources) and dangerous solution.
3. IPv6 addresses availability makes possible to use a global address for every device in the home network. This IPv6 address can be sent to other control points in remote networks without the need to create redirections in the residential gateways.

Using SIP and IPv6 the devices from a residence will be seen in the Control Point of the other one because they have a public address accessible from any other residence.

The first step to make SIP work is to have a SIP Proxy working. This proxy is needed so that Control Points can register themselves and be localizable and exchange devices. The SIP Proxy used in this work is a modified version of the Open Source Java implementation of the National Institute of Standard and Technologies (NIST) SIP Project [26].

Following the SIP way to work both Control Points must be registered to the SIP Proxy as seen in Fig. 4.5. Once the Control Points are registered the Home A user sends the devices to Home B user so that he can use the locally the Media Server (see Fig. 4.6).

Once the UPnP devices are sent the user at Home B has all the needed devices render content from the remote network: the own renderer and the media server of the other home, as seen in Fig. 4.7.

When the *Home B Control Point* has the *Home A Media Server* the user can the wished play content from that server.

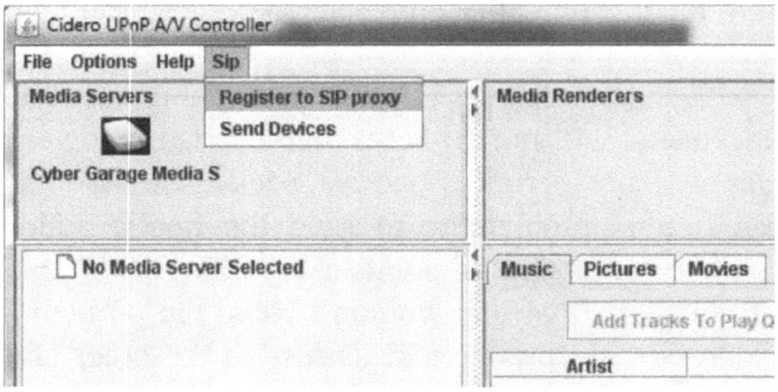

Fig. 4.5 The home A user registers the control point in the SIP proxy

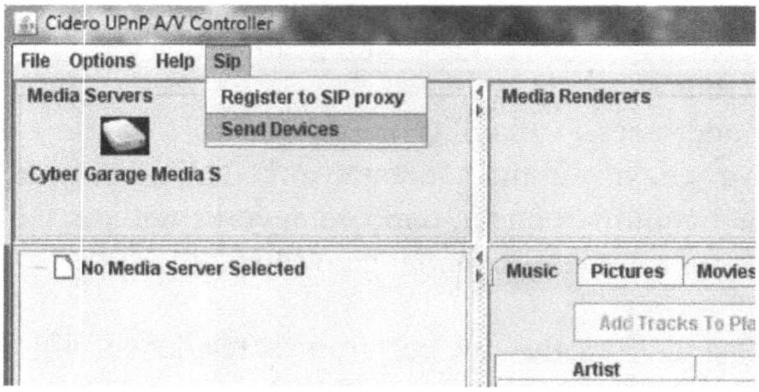

Fig. 4.6 Home A: user sends devices using the control point

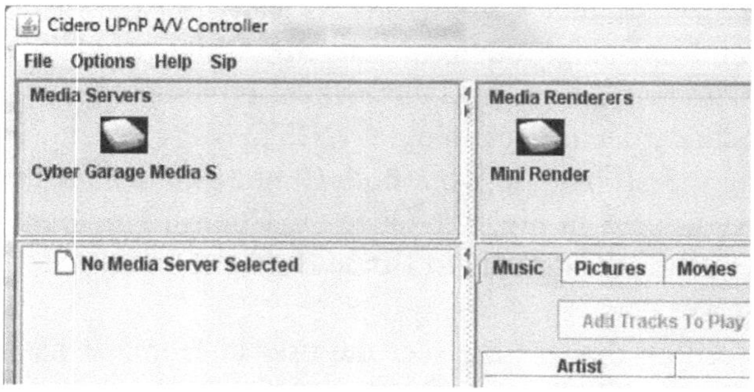

Fig. 4.7 Home B: control point with remote media server and local renderer

4.6 Conclusion

In this work it has been shown how, with the combination of SIP and UPnP, a delocalized use of media content can be easily done under a OSGi framework getting the management aspect. It is crucial since it maintains the whole life cycle of services for the user.

The use of UPnP end to end opens the possibility to delocalize specific content while the only important issue is where it is going to be reproduced.

The migration to IPv6 should arrive in 2010 as several reports [27] suggest since time ago. But configuring a stable environment for the scenario required a bigger effort than the expected. It didn't only require the modification of Software applications like Cidero, Cyberlink and MediaGate, but also de modification of the firmware in the hardware used like routers for them to properly work in IPv6.

However the big quantity of devices with high technical capabilities like network connection and multimedia functionality makes delocalization of services mandatory. The use of IPv4 not only complicates the problem (i.e. using NAT). It supposes a great delay to provide solutions for a technology to be substituted relatively soon, something that happens with most of the actual developments. This work has been done multiplatform to get the needed versatility and keep transparency for the user.

As last it hast to be mentioned that the end to end model in media content sharing opens new possibilities not only to user level but also to service providers who could provide a new business model offering content to a delocalized client.

Acknowledgments This work has been partly developed inside the PLANETS project (MEDEA+ A-306), financed by the Spanish Ministry of Industry under project number FIT-330220-2005-111.

References

1. Session Initiation Protocol (SIP) Forum, April 2008. URL: http://www.sipforum.org
2. Universal Plug and Play (UPnP) Forum, April 2008.URL: www.upnp.org
3. Universal Plug and Play (UPnP) UPnP AV Architecture:1, June 2002 URL: www.upnp.org/specs/av/UPnP-av-AVArchitecture-v1-20020622.pdf
4. Internet Protocol Version 6, IPv6, April 2008. URL: http://www.ipv6.org/
5. A. Vallejo, J. Ruiz, J. Abella, A. Zaballos, J.M. Selga "State of the Art of IPv6 Conformance and Interoperability Testing", IEEE Communications, October 2007.

6. A.E. Nikolaidis, S. Papastefanos, G.A. Doumenis, G.I. Stassinopoulos, M.P.K. Drakos, "Local and Remote Management Integration for Flexible Service Provisioning to the Home", IEEE Communications Magazine, October 2007.
7. M. Ditze, I. Jahnich, "Towards End-to-End QoS in Service Oriented Architectures" in Industrial Informatics INDIN, 2005.
8. A. Vilei, G. Convertino, F. Crudo, "A New UPnP Architecture for Distributed Video Voice over IP", in Proceedings of the 5th International Conference on Mobile and Ubiquitous Multimedia MUM'06. December 2006.
9. Y. Nakamoto, N. Kuri, "Siphnos – Redesigning a Home Networking System with SIP", in Proceedings of the 6th IEEE International Conference on Computer and Information Technology, 2006.
10. A. Brown, M. Kolberg, D. Bushmitch, G. Lomako, M. Ma, "A SIP-based OSGi Device Communication Service for Mobile Personal Area Networks", in IEEE CCNC 2006 proceedings.
11. Open Service Gateway Initiative (OSGi) Alliance, 2008, URL: http://www.osgi.org
12. P. Dobrev, D. Famolari, C. Kurzkey B.A. Miller, "Device and Service Discovery in Home Networks with OSGi", en IEEE Communications Magazine, August 2002.
13. Domoware, UPnP Bundles for OSGi. WWW, 2006. URL: https://sourceforge.net/project/showfiles.php?group_id =118919id=118919
14. D.-O. Kang, K. Kang, S.-G. Choi, J. Lee, "UPnP AV Architectural Multimedia System with a Home Gateway Powered by the OSGi Platform", in Proceedings of the IEEE International Conference in Consumer Electronics, January 2005.
15. Universal Plug and Play (UPnP) Forum, MediaServer V 2.0 and MediaRenderer V 2.0, March 2006. URL: http://www.upnp.org/specs/av/
16. Dinamic Dns, DynDNS, April 2008.URL: http://www.dyndns.com/
17. P. Eronen, IKEv2 Mobility and Multihoming Protocol (MOBIKE), Internet Engineering Task Force, RFC 4555 (Proposed Standard) (June 2006)
18. Personal domain of Satoshi konno, , 2007. URL: http://www.cybergarage.org
19. UPnP Intel Implementation. WWW, 2007. URL :http://www.intel.com/cd/ids/developer/asmo-na/eng/downloads/upnp/overview/index.htm
20. Cidero Software Solutions for the Digital Home, WWW, 2007. http://www.cidero.com
21. Media Streamer for Nokia 770 Internet Tablet, WWW, 2007. http://www.nseries.com/770experience_2/hacks_streamer.htm
22. Microsoft, Windows Media Player 11, 2007. http://www.microsoft.com/windows/windowsmedia/es/player/11/default.aspx
23. UPnP Media Render implementation for Linux. WWW, 2007. http://soggie.soti.org/gmediarender/
24. Java Media Framework, April 2008, URL:http://java.sun.com/products/java-media/jmf/
25. Jmf Mp3 Plugin, April 2008, URL: http://java.sun.com/products/java-media/jmf/mp3/download.html
26. Jain-SIP-Presence-Proxy, April 2008, URL:https://jain-sip-presence-proxy.dev.java.net/
27. The great IP crunch in 2010,Mark Frauenfelder, September 21 1999, http://www.cnn.com/TECH/computing/9909/21/ip.crunch.idg/index.html

Chapter 5

Virtual Development Environment for Embedded Systems Using ARMulator and SystemC Models

Sang-Young Cho[1] and Jeong-Bae Lee[2]

[1]*Computer Science & Information Communications Engineering Division, Hankuk University of Foreign Studies, Yongin, Kyeonggi, Korea;* [2]*Computer Information Department, Sunmoon University, Asan, Chungnam, Korea, sycho@hufs.ac.kr*

Abstract Virtual development environment increases efficiency of embedded system development because it enables developers to develop, execute, and verify an embedded system without real target hardware. This chapter deals with an implementation of a virtual development environment for ARM core-based embedded systems. The environment is based on ARM's ARMulator simulation environment and extended to use SystemC models by attaching a SystemC simulation engine to the ARMulator. Therefore, the environment can flexibly use both ARMulator-based and SystemC-based hardware models. We developed some hardware IP modules and user interface programs to enrich the environment for hand-held devices or general application development. In addition, a real-time operating system μC/OS-II is ported on the environment so that it can be used to develop multi-thread applications. Compared to other environments, its construction cost is very low and the environment can be easily modified according to an engineer's needs.

Keywords Virtual development environment, ARMulator, SystemC, Embedded system Development, Simulation.

5.1 Introduction

On-time delivering of an embedded system solution to market with the complete required functionality is very crucial because the market is highly competitive and the demands of consumers rapidly change. Most embedded systems are made up of a variety of Intellectual Property (IP) including hardware IP's (the processors and peripherals) as well as software IP's

N. Martínez Madrid, R.E.D. Seepold (eds.), *Intelligent Technical Systems*, Lecture Notes in Electrical Engineering 38, DOI 978-1-4020-9823-9_5,

(operating systems, device drivers and middleware like protocol stack). In the traditional development flow, application software design does not start until after these IP's are integrated or nearly integrated. Thus, the resulting development cycle is too long for the competitive market. The ideal solution must permit the software development team to perform its tasks before the hardware or a physical prototype is available so that the critical time-to-market requirements could be satisfied. Virtual Development Environment (VDE) is virtually built inside a computer, and it provides the same appearance as real one and simulates various peripherals attached outside of the board and functions of them as well [1].

Virtual Development Environment is used to verify a hardware prototype, develop software without the real hardware, or be used for co-design of hardware and software for embedded systems. This environment usually provides a hardware simulation model, simulation engine, and other tools that are useful in software development. Thus it increases the efficiency of the embedded system development [2,3,4]. Though it does not provide with the same appearance of the target system, application software could be developed and fully verified using hardware or software IP models.

Virtual Platform [2] is Synopsys's commercial virtual development environment. It supports many different types of processors such as ARM, X-Scale, and MIPS that are usually used in embedded systems. Virtual Platform supports to develop software for target systems. RealView Soc Designer [3] is comprehensive SystemC-based SoC development environment provided by ARM. It consists of fast and easy modeling and simulation, and debugging tools. Additionally, it enables a system or hardware developer to compose the most suitable architecture quickly and accurately, and software developers to develop software before the actual hardware comes out by providing VDE. Visual Elite [4], made by Mentor Graphics, is a set of 'Fast ISS' models and 'platform-based packages', which are composed of TLM-based busses, memory, and peripheral device models. According to its setting, it can comprise and run many different abstract-level systems, and debug-related software and firmware. Also, it can create a virtual environment for fast software development. The commercial VDE's described above are tightly integrating the hardware simulation tools and software development tools. Therefore, it limits the flexible use of various software tools and hardware simulation tools from different companies or organizations.

In this chapter, we describe the design and implementation of a virtual development environment for developing embedded systems that use ARM processor cores. We construct a virtual development environment based on ARMulator that is an ARM core instruction set simulator of ARM cooperation. The ARMulator environment uses a specific hardware modeling method. We modeled the ASB bus that is an in-chip bus scheme developed by ARM and we extended the environment by attaching several hardware IP's. Also, the developed environment is extended to use SystemC hardware IP's by attaching a SystemC simulation engine to the modeled ASB bus. Therefore, the environment can use both ARMulator-based hardware models and SystemC-based hardware models. Especially, the environment is constructed with ARM920T core that equips with cache, MMU, ASB interface, etc. We built the virtual target hardware environment that is composed of various ARMulator or SystemC hardware models such as Memory controller, LCD controller, Interrupt controller, 1-ch DMA, UART, 2-ch Timer, Watchdog Timer, and GPIO Ports.

Actually, a software development can be accelerated by various behavior viewers of hardware IP's rather than by a real appearance of a target system. We implemented several graphical user interfaces such as LCD panel, Timer viewer, UART interface, and LED display. The implemented VDE can be used to develop an LCD-based hand-held equipment. And, a real-time operating system μC/OS-II is ported on the virtual environment. Thus it can be also used to develop μC/OS-II-based applications.

The rest of the chapter is organized as follows. In Section 5.2, we describe the ARMulator environment and SystemC that are necessary for developing the environment. In Section 5.3, we explain the design and implementation of our virtual development environment for mixed model and multi-threaded application development environment. Section 5.4 describes the verification methods and results of the hardware IP modules and μC/OS-II operations. Finally, Section 5.5 concludes the chapter with final remarks.

5.2 Related Studies

5.2.1 ARMulator Environment

For the processor vehicle of a virtual software development environment, ARM920T core is selected due to its popularity. The most of hand-held terminal equipments are adopting ARM cores as their main control processors [5]. We also choose the ARMulator environment as the base one of our virtual software development environment because it is too time-consuming to build our own various ARM processor core simulators and to manage to follow up the rapidly-announced ARM core series. Furthermore, most of software developers for ARM systems have an ARM's software development environment such as ADS (ARM Developer Suite) and RVDS (Realview Developer Suite) series and ARMulator is one of components of the ADS or RVDS series [6,7]. An ARMulator-based virtual software development environment may enable developers to have a virtual environment with a minimum cost.

ARMulator is a virtual board model that stands on the basis of cycle-based instruction set simulator of ARM processor cores. As it has a simple memory model and a standard hardware IP, the performance of hardware and software can be estimated before the actual implementation of the product by using the software's profiling function even if no tangible board exists [8,9]. Figure 5.1 shows the ARMulator framework.

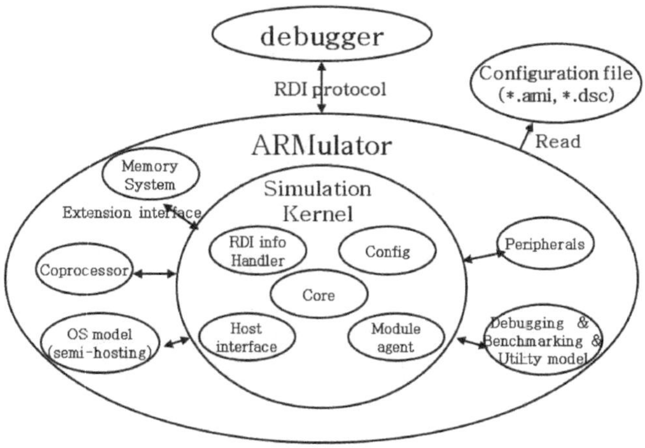

Fig. 5.1 ARMulator DLL structure

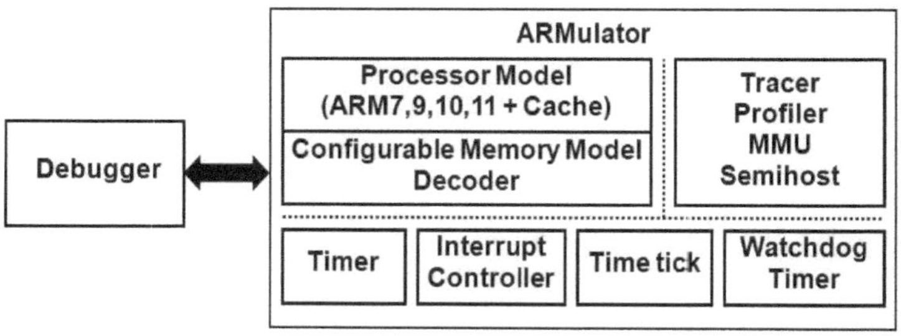

Fig. 5.2 Basic ARMulator environment

ARMulator consists of simulation kernel and extension modules. Simulation kernel is in charge of core simulation and external interface with RDI debuggers or extension modules. Extension modules include basic peripherals such as timer, watchdog timer, time tick, and interrupt controller. Even though software development is possible with those peripherals as well as core simulator in the ARMulator virtual environment, it is far from a real environment where various peripherals are working. ARMulator is open to expand peripherals with the extension interface of memory system. Figure 5.2 shows basic virtual software development environment of ARMulator.

5.2.2 SystemC

SystemC is a system modeling language that can model and operate hardware at system-level. SystemC can easily express a complex SoC core at a high level while having all the merits of the hardware description languages. It was developed using C++ classes. Hence, SystemC can effectively be used for simulation environment in checking not only hardware but also software operations. Also it supports TLM(Transaction-Level Modeling) [10,11].

We implemented a virtual development environment using SystemC version 2.0.1. Version 2.0.1 provides not only register-level transmission modeling, but also algorithm-and-function-level modeling. SystemC class libraries provide essential classes for modeling system structure. They support hardware timing, concurrency, and reaction, which are not included in standard C++. SystemC allows developers to describe hardware and software, and their interface under the C++ environment. Main parts of SystemC are composed as follows.

- Module: A container class that can include other modules or processes.
- Process: Used for modeling functionality and defined within a module.
- Ports and Signals: Signals connect modules through ports. (Modules have ports, through which they are connected to other modules.)
- Clock: SystemC's special signal. (act as a system clock during simulation.)
- Cycle-based simulation: Supports an untimed model and includes a high-level function model to RTL-level, which has clock cycle accuracy.

Figure 5.3 shows an example of a system modeling in SystemC. A module can include processes or other modules due to its hierarchical structure. Processes run concurrently and do a function modeling. Thus, they cannot have a hierarchical structure. Processes can communicate with each other through channels. Signals are modeled in the simplest form of the channels.

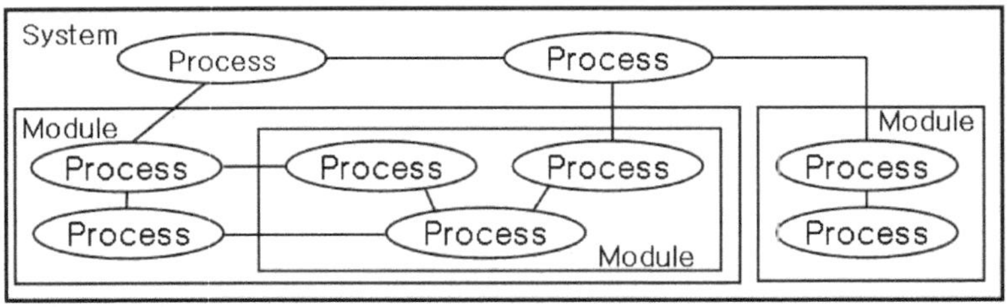

Fig. 5.3 A system modeling of SystemC

5.3 Design and Implementation

5.3.1 Extension of ARMulator Environment

ARMulator can be extended by adding any hardware that is modeled in C language as a DLL file. The developed environment also can be extended with SystemC hardware models. Figure 5.4 shows the target system hardware modules to be added to form a virtual development environment.

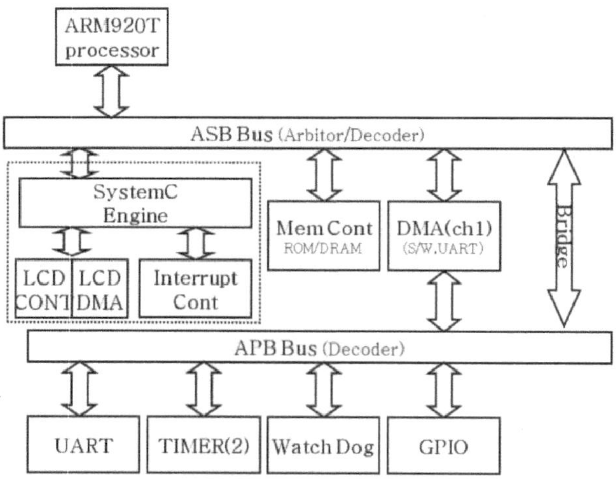

Fig. 5.4 Target system modules

The target system modules are designed and implemented by referencing to S3C2410 hardware IPs. S3C2410 made by Samsung electronics is an SoC chip for hand-held devices and general applications [12]. The ARM920T IIS simulator of ARMulator is used as the CPU core of the target system. The

ARM920T implements MMU, AMBA BUS, and Harvard cache architecture with separate 16 KB instruction and 16KB data caches, each with an 8-word line length. The other modules include Memory controller, LCD controller, Interrupt controller, 1-ch DMA, UART, 2-ch Timer, Watchdog Timer, and GPIO Ports. We modified SystemC engine module to be connected with the designed ASB Bus interface.

Via the SystemC engine module, an arbitrary SystemC hardware model can be easily connected into the virtual environment. Currently, an LCD controller and an interrupt controller are implemented in our environment. The other hardware IP modules are developed by using the ARMulator's module development APIs. The operation of modules is similar to that of hardware IPs of S3C2410. Most chips from Samsung use the same hardware IP of S3C2410. Therefore, the constructed virtual environment has many benefits for developing software with Samsung chips. ASB and APB buses are used to interconnect modules and there is a bridge block between two buses. The two buses are implemented to ensure functionality and transfer time rather than an actual transfer protocol. A priority-based arbiter is implemented because there are three masters connected on the ASB bus.

5.3.2 SystemC Extension

To use SystemC hardware models in the ARMulator environment, we connected a simulation engine of SystemC version 2.0.1 to the Armul_bus module in Flat Memory of ARMulator. The original Armul_bus in the ARMulator environment enables various hardware IP's to be accessed by the processor core. We modified the Armul_bus to be capable of ASB operations. Figure 5.5 shows the overall structure of the developed SystemC connection environment.

If we set AxD (Debug controller) [6] to connect the implemented VDE, system functions related to a connection are called and finally Armul_bus initializes SystemC modules via Csimul class. SystemC generally uses SystemC.lib for modeling and simulation. Therefore, to connect Armul_bus, we analyzed the internal procedure of SystemC.lib and modified it according to our needs. To do this, we removed main() from the SystemC.lib, which starts a simulation, and re-built the SystemC.lib. Then we connected the starting function, sc_main(), of the SystemC.lib with the InitializeModule() of the Armul_bus. Thus, the simulation can be controlled by InitializeModule(). The class Csimul is implemented for the connection of the Armul_bus and the SystemC simulation modules.

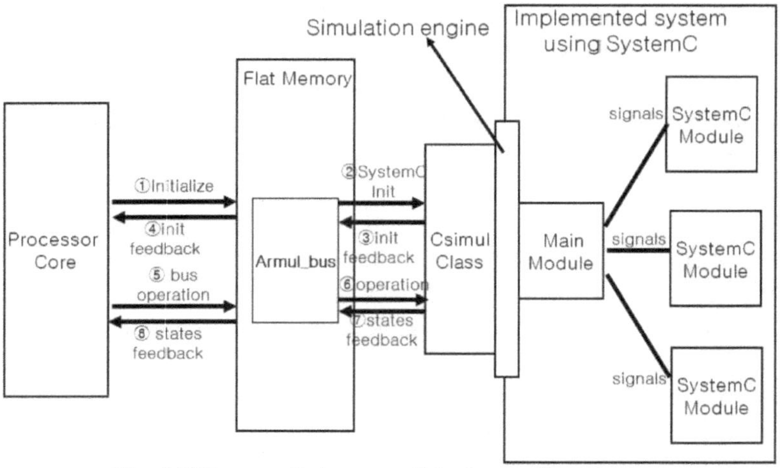

Fig. 5.5 The overall structure of the SystemC connection

The behavior of class CSimul is as follows:

1. Make sc_signal for input/output wires of modules.
2. Connect signals after a main module in SystemC is made.
3. Create functions necessary for reading and writing a state of the connected modules.
4. During simulation, a callback function is called by Armul_bus.
5. Allow results of simulation to be reported to Armul_bus.

After initialization is set up, a SystemC run environment is created and Armul_bus recognizes SystemC-based IP modules according to its setting. A simulation can start when all the necessary modules are created and the connection between signals and modules are validated. Armul_bus calls CSimul's method, CSimul.init(), to control a starting of a simulation. This initialization process is to start the most top module of a simulation, and an actual simulation starts when sc_start() is called from the most top phase. The function sc_first() can have a double-type parameter and various time units as its value. If we want to run a simulation without a time limit, we can put a negative number to the parameter. All these functions run while the clock signal ticks for an assigned amount of units. When the time unit is all spent, a SystemC scheduler is called.

A simulation can be terminated at any time when sc_stop() is called with no parameters. But it is difficult to understand all the details above and to implement exact operations of each bus cycle step that the Armul_bus requires. To solve this problem, we used sc_initialize() function that starts a clock and controls a simulation, rather than using sc_start(), which SystemC provides.

We initialize the SystemC scheduler using the sc_initialize() function when the Armul_bus InitializeModule() function is called. Then, we will be able to write values on the signals and simulate results of the set values. This function takes a double-type parameter and can call the SystemC scheduler. To implement one-step operation, we used the sc_cycle() function and in this way SystemC modules are synchronized with the ASB operation clock.

5.3.3 Peripheral Features of the Implemented Environment

For a virtual development environment for ARM-based equipments, some peripheral features are implemented to verify functions and behaviors of extended modules. An independent Windows application is programmed to mimic an LCD panel and LED dispaly. Actually, the software development process can be accelerated by various behavior viewers of hardware IPs rather than the real appearance of a target system. A SystemC model LCD controller is virtually connected via Windows IPC (Inter-Process Communication) with the LCD panel that displays image data from the LCD controller modules. Some GPIO ports are virtually connected with 4 LED's and the on-off control of GPIO ports are captured on the LED graphics of the program. UART modules use device control Windows APIs so that a real COM port can be used to verify UART module's behavior. If the UART module of a virtual target system is programmed to send or receive data, it will appear at a real serial port and it can communicate with other real hard serial ports. The implemented timer can be programmed to generate PWM (Pulse Width Modulation) signal. For now, the trajectory of the generated PWM signal is stored in a log file to be verified later. Figure 5.6 shows the constructed virtual development environment.

To run a program in the virtual environment, an execution image created by the ARM software development tool should be downloaded to memory, and debugged by ARM's debugger. An LCD-based application can be easily developed in the environment because it has sufficient hardware modules for LCD operations. To run an LCD-based program, the LCD controller needs to be programmed appropriately according to the LCD panel type, color representation format, image buffer location in memory. After the LCD controller starts, the image buffer for the LCD panel should be filled with pixel data so that the LCD controller DMA transfers the pixel data to the LCD panel periodically. As the environment contains hardware IP modules that are used frequently, a single threaded application can be easily run on the virtual environment. The developed environment can be easily expanded or changed by adding or replacing ARMulator-based or SystemC-based hardware IPs.

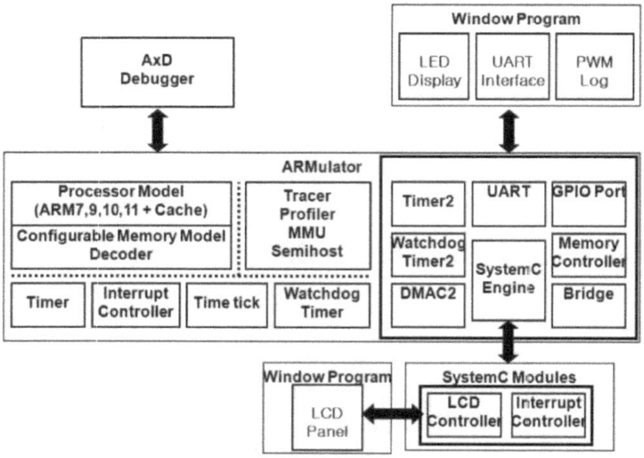

Fig. 5.6 The developed virtual development environment

5.3.4 Porting μC/OS-II on the Virtual Environment

Embedded applications can be classified into two categories: single-threaded application or multi-threaded application. Usually, a deeply embedded application that performs a simple monitoring and a dedicated control is constructed as a single-threaded application. When a system should perform a lot of tasks concurrently, a system is built as a multi-threaded system with the help of an embedded operating system. We expanded the implemented virtual development environment to be used as a multi-threaded application development platform by porting μC/OS-II on it. The real-time kernel μC/OS-II is a portable, ROMable, scalable, preemptive real time and multitasking kernel for microprocessors and microcontrollers [13]. μC/OS-II can manage up to 63 application tasks and provide the following services: semaphores, event flags, mutual exclusion semaphores to reduce priority inversions, message mailboxes, message queues, task management, fixed sized memory block management, and time management. Due to its simple and compact features, μC/OS-II is used for many commercial and educational purposes. μC/OS-II is layered into hardware-dependent part, hardware-independent part, and application-dependent part. Among three parts, the hardware-dependent part is the most considerable when porting takes place. A processor core, a timer and an interrupt controller are essential hardware IP's when files in hardware-dependent part are modified for porting. The files (OS_CPU.H, OS_CPU_A.S, and OS_CPU.C) in hardware-dependent parts are modified according to hardware control features of the implemented target system modules. Figure 5.7 shows the μC/OS-II ported on the virtual environment structure.

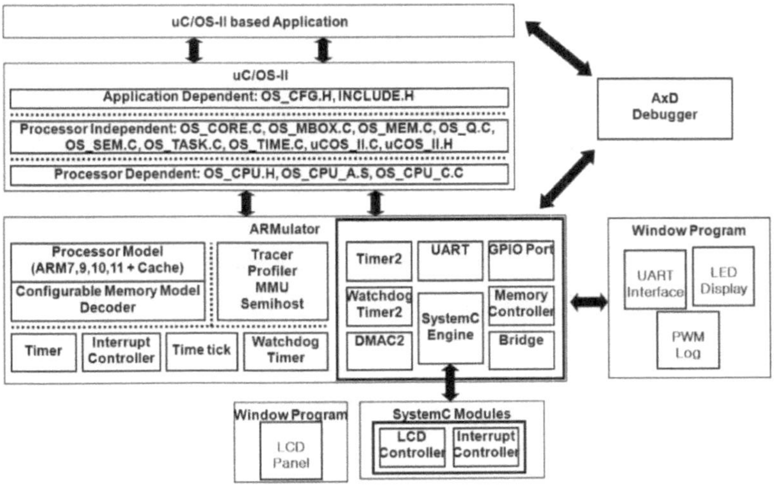

Fig. 5.7 OS-based virtual environment structure

In this environment, we can develop not only a single-threaded application but also a multi-threaded application that use the μC/OS-II APIs.

5.4 Verification of the Implemented Environment

5.4.1 Verification of Hardware IP Models

To verify the implemented virtual development environment, some test programs for hardware IP's are programmed. A simple digital clock program is programmed to verify functions of Timer, Interrupt controller, LCD panel, and LED display. The program displays the changes of time/minute/second using the operation of Timer and Interrupt controller and shows the On-Off behavior of LEDs. UART is verified with a program that sends character strings to an external computer connected with the host computer of the implemented virtual development environment through a COM port. Also, a program is written to control Timer model to generate various PWM waves and the changes are logged in a log data file. The viewer displays the PWM wave from the log data file.

5.4.2 Verification of μC/OS-II Operations

A simple three-task application was programmed to verify the ported μC/OS-II operations. The most important function of an operating system is task scheduling and correct operations of timer and interrupt controller are required to implement the function.

During the test application running, three tasks with different priorities were created to test a scheduling. The main function creates a task (TASK1), and TASK1 creates two tasks (TASK2 and TASK3). Each task draws its own position on the LCD panel by writing its image data to image buffer in the main memory. The position of each task is moving side-to-side one pixel at a time with different delay values in an infinite loop. TASK1 has 100 Ticks delay, TASK2 has 200 Ticks and pends a semaphore, and TASK3 has 400 Ticks and posts a semaphore. The program runs without any error on the multi-threaded environment.

The operation of the program confirms that of the LCD controller and the LCD panel. And it also verifies the scheduling function of μC/OS-II that uses timer and interrupt controller.

5.5 Conclusion

A virtual development environment based on ARMulator is designed and implemented to support a development team to perform hardware/software tasks before a target hardware or a physical prototype is available. A virtual development environment reduces the cost of developing an embedded system by enabling engineers to write embedded software without real hardware. The implemented environment supports not only ARMulator but also SystemC hardware models. To achieve this, we attached a SystemC simulation engine to the designed ASB bus. We minimized a modification of SystemC simulation engine so that the environment can be easily changed or extended with various SystemC models.

The environment provides a virtual hardware platform that is equipped with ARM920T processor core, hardware IP modules, and graphical peripheral user interfaces. The current processor core can be easily changed with the other supported core of ARMulator. The ARM processor cores are the most commonly used core in commercial business [5].

The hardware IP modules are designed and developed referring to Samsung's hardware IPs. The graphical user interfaces for peripherals such as an LCD panel and LED's can help developers to verify their program behavior. The UART IP module is implemented so that it can communicate with real terminal equipments. With this environment, a developer can design, program, and verify a single-threaded application as like he/she works on a real hardware-based prototype. Using the popular software development tool such as ADS or RVDS series, the environment can be constructed with a minimum cost. We expanded the implemented virtual software development environment to be used as a multi-threaded application development platform by porting μC/OS-II on the virtual environment. A simple three-task application was programmed to verify the operation of the implemented virtual hardware IPs of the target system and the ported μC/OS-II operations. Software design and development for embedded system using LCD can benefit from the implemented environment, and μC/OS-II-based application program can be developed and run on the virtual environment. The developed virtual development environment can be used in many phases of embedded software development such as developing a device driver, porting an OS, and developing an application.

For further works, more hardware IP modules related to sounds, serial communication, etc. should be supplemented. Individual hardware IP modules should be able to be reconstructed to support various embedded system development environment. Moreover, various graphical user interface modules should be implemented to enable developers to develop software in a convenient virtual software development environment.

Acknowledgments This research was supported by the MKE (Ministry of Knowledge Economy), Korea, under the ITRC (Information Technology Research Center) Support program supervised by the IITA(Institute of Information Technology Advancement) (IITA-2008-C1090-0801-0020).

References

1. T. Anderson, R. Schutten, and F. Thoen, Virtual Prototypes Cut software Bottleneck, Wireless System Design, http://www.wsdmag.com/Articles/ArticleID/9821/9821.html, February 2005.
2. Synopsis Corp., VPDA295 Virtual Platform, http://www.virtop.com/products/page/0,2573,33,00.html, 2008.
3. ARM Corp., Virtual Prototyping Solution, http://www.arm.com/products/DevTools/RealViewSystemDevelopment.html, 2008.
4. Mentor Grpahics Corp., Platform based Solutions, http://www.mentor.com/products/esl/system_integration/visual_elite/index.cfm, 2008.
5. W. East, ARM Holdings plc Morgan Stanley-7th Annual TMT Conference, November 2007.
6. ARM Cop., ARM Developer Suite 1.2 Debug Target Guide, November 2001.
7. ARM Cop., Realview ARMulator ISS User Guide, January 2004.

8. S. Furber, ARM System on chip Architecture 2/E, ISBN 0-201-67519-6, Addison Wesley, 2000.
9. ARM Cop., ARM DAI0032E Application Note32: The ARMulator, Sep., 2003.
10. T. Grotker, System Design With SystemC, Kluwer Academic Publishers, 2002.
11. F. Ghenassia, Transaction Level Modeling With SystemC, Springer Verlag, 2006.
12. Samsung Electronics, S3C2410A User's Manual Revison 1.0, March 2004.
13. J.J. Labrosse, MicroC/OS-II Real Time Kernel 2/E, ISBN 1-578-20103-9, R&D Technical Books, 2002.

Part II

Embedded Programming

Chapter 6

Rule-Set Extraction from C-Code

Franz Wotawa and Willibald Krenn
Institute for Software Technology, Technische Universität Graz, Inffeldgasse 16b/2, A-8010 Graz, Austria, wkrenn@ist.tugraz.at

Abstract We present an approach for extracting knowledge from C source code of control programs. The extracted knowledge is intended to be used in our smart control engine which takes a rule set and decides which rules to use based on the internal and environmental conditions. The extraction of rules is based on the control-flow graph of the supplied C program: Basically, our method extracts rules that correspond to paths to given high-level function calls. The advantage of this method is to get a first knowledge-base from available source code which makes using a smart control engine more applicable for industry. We use an industrial control program as example within the paper in order to justify the usefulness of our approach.

Keywords Control Flow Graph; Conversion Problem; Abstraction; Program Transformation

6.1 Introduction

Autonomous and mobile devices have special needs with respect to reliability and the capability to react and adapt on changed environmental and internal conditions. For example, a device which is attached to a container for the purpose of sending position information of the container during a travel from one location to another should be capable to deal with faults like missing mobile connections or running out of power. Even in cases where the whole functionality cannot be ensured anymore, basic capabilities which are important to fulfill a certain mission should be retained as long as possible.

In order to implement such smart behavior there are two possibilities available. First, someone might create a control program where all future

N. Martínez Madrid, R.E.D. Seepold (eds.), *Intelligent Technical Systems*, Lecture Notes in Electrical Engineering 38, DOI 978-1-4020-9823-9_6,

interactions have been foreseen in advance which is very unlikely especially for more complex systems and the environments they are interacting with. Second, we enable the device to reason about its capabilities. This requires us to come up with a knowledge base that captures possible behaviors and let the device autonomously select the one that is best in a certain situation. In addition, preference criterions for behaviors can be used to control the selection of the behavior to some degree.

In this chapter, we assume that we have a smart control engine [1] which relies on a knowledge base for controlling the device behavior. In particular we assume that the knowledge base comprises rules which are selected and executed based on external events and the internal state. In case of faults the selection of the rules is adapted accordingly. In this adaptation process someone might use a heuristics which ensures that those rules which are more likely to be executed without any problems are more often used without preventing rules to be never executed. The focus of this paper is not on the smart control engine but on the extraction of the required knowledge from available control programs. Two observations motivated us for the research behind the paper:

- People in industry are very experienced in writing control programs but less experienced in developing knowledge bases.
- Control programs are available and have usually a very similar structure.

Hence, by offering a technology for extracting high-level rules, that form our knowledge base, from control programs, we are building a transition path from conventional to intelligent systems that respects and re-uses existing investments, including existing knowledge. In fact the main purpose of the presented approach is to extend existing control programs: We aim at the re-use of existing low-level parts of the code, while replacing high-level parts with the smart control engine.

It follows that the behavior of the original control program and the behavior of the smart control engine using the extracted knowledge base have to be as similar as possible. Similarity in this context means that the extracted knowledge together with the smart runtime should compute the same output as the original program using the same input values.

We do not expect the extraction process to be perfect in a sense that no manual changes to the extracted knowledge base are necessary. Hence, after extracting knowledge from the program we allow for a process which changes the obtained rule set in order to optimize the behavior of the smart control engine.

The proposed conversion of control programs into their rule based representation assumes that the control program implements some sort of finite automaton. In such an implementation an infinite control loop comprises conditional statements which implement the state changes of the corresponding automaton. The underlying idea of the conversion is to extract rules which correspond to execution paths within the infinite control loop. Such execution paths are equivalent to the paths from the start state of the automaton to a state which is connected with the start state.

The paper is organized as follows. In the next section we describe the problem in detail. For this purpose we use a simple control program and discuss our underlying smart control engine. Afterwards we formalize the basic conversion process and apply it to the control program. Next, we present results gained from a tool that implements the presented approach and discuss limitations of our approach. Before finally concluding the paper, we give an overview of related research.

6.2 Basic Idea

In this section we introduce the problem of extracting knowledge from C programs and our proposed solution. For this purpose we use a running example which is partially given in Fig. 6.1. The whole program controls a mobile device which tracks its position and periodically sends it to a server using a GSM communication line. The device also allows for receiving messages from the server. One idea represented in the code is only to send the position if it has changed. This idea is implemented in lines 17-24 using the signal quality of the GSM connection as indicator for movement.

When extracting knowledge from such a program we are interested in identifying parts of the program which should remain unchanged because they provide a behavior which maps from a high level function to low level software which is directly connected with the underlying hardware. Hence, we have to have knowledge about the high level functionality we want to extract. In our case we are interested in extracting rules which correspond to the conditionals occurring within the control loop (lines 4–6) which are basically constructed from conditions and functions called. Since, functions or procedures might call other functions and procedures we have to know when to stop the conversion. We do this by assuming that we know functions or procedures which are seen as being atomic components for the conversion. We are not going to look inside such components for the purpose of knowledge representation.

In our example we are interested in gaining knowledge which corresponds to the functions *makePassiveCall* and *makeActiveCall*. Using this information we extract the execution paths leading to the execution of the given functions.

```
1  void main (void)
   {
3          /* ... */
           while (1) {
5                  runM2M();
           }
7  }
   /* ... */
9  void runM2M(void)
   {
11         if (gsm_ev_ring() == 1) {
                   makePassiveCall();
13                 setM2MReportTimer(REPORT_TIMEOUT);
           }
15         if (getM2MReportTimer() == 0) {
                   setM2MReportTimer(REPORT_TIMEOUT);
17                 if (getGPSState() == 'S') {
                           char old, new;
19                         old = getSignalQuality();
                           gsm_act_readdb();
21                         m2m_wait(3);
                           new = getSignalQuality();
23                         if (new == old)
                                   return;
25                 }
                   makeActiveCall();
27         }
   }
```

Fig. 6.1 Excerpt from a control program written in C

These execution paths can be represented by the conditions which have to be evaluated to true (or the negations of the conditions to evaluate to true) so that *makePassiveCall* or *makeActiveCall* is called. In our example, we know that *gsm_ev_ring()==1* has to evaluate to true in order to reach the function call *makePassiveCall()*. We now represent this information as a rule of the form $cond_1 \rightarrow passiveCall$ where $cond_1$ stands for *gsm_ev_ring()==1* and $passiveCall$ for the statements in line 12 and 13. The semantics behind such a rule is that if $cond_1$ is true, $passiveCall$ is true which leads to the execution of statements 12 and 13. Hence, $passiveCall$ can be seen as a function that is going to be executed whenever $cond_1$ evaluates to true.

In order to get a rule for *makeActiveCall()* in line 25 we use an extension to the idea already discussed. The problem here is that statements of line 17–24 have an influence on the execution of line 25. Depending on the evaluation of *new==old* in line 23 *makeActiveCall()* is executed or not. However, both variables *new* and *old* are local variable. Hence, their values have only local influence. The conditional expression used in the other if-then-else statements does not use local variables which makes the transformation into rules much easier. A solution to the problem of local variables in the conditional is to generate a new function *doCheck()* which comprises the code of lines 18–24. This function returns true if line 23 evaluates to true and false, otherwise. This change preserves the behavior of the program.

```
int doCheck() {
    char old, new;
    old = getSignalQuality();
    gsm_act_readdb();
    m2m_wait(3);
    new = getSignalQuality();
    if (new == old)
        return 1;
    else
        return 0;
}
```

Using this function, lines 17 to 25 can be re-written as follows:

```
if (getGPSState() == 'S')
    if (doCheck() == 1)
        return;
makeActiveCall();
```

Hence, finally we are able to extract a rule $cond_2 \wedge (\neg cond_3 \vee \neg cond_4) \rightarrow activeCall$ where $cond_2$ corresponds to *getM2MReportTimer() == 0*, $cond_3$ to *getGPSState() == 'S'*, and $cond_4$ to *doCheck() == 1*. The two rules together with their corresponding low-level code now can be used to control the device. Before they can be loaded into our smart control engine, some small additions have to be made: We need to insert a rule that tells the execution engine that it is good for the system to execute one of the extracted rules. This is done by adding a goal that says that it is desirable to run any of the two rules. In case a rule can be executed, the corresponding low-level functions that were preserved by importing the original source code will be called by the smart control engine.

In the following section we introduce a way of automatically extracting a rule set from the source code.

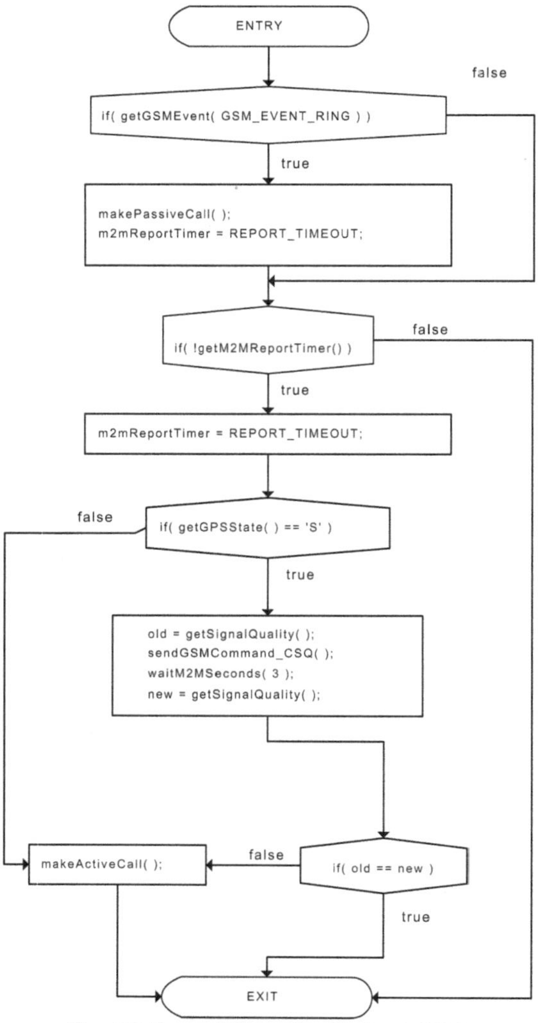

Fig. 6.2 Control Flow Graph of runM2M

6.3 Conversion Process

Starting from the underlying idea behind the conversion which has been explained in the previous section, we now are going to formalize the whole process. The formalization of our solution is based on the concept of the well-known control flow graph (CFG) which captures the flow of the control through a program. Based on the CFG we define functions for extracting the required information and for changing the CFG whenever appropriate for extraction.

Definition 1 (Control Flow Graph) *A control flow graph (CFG) is a tuple* (V, A) *where* V *is the set of vertices where each vertex represents a basic block, i.e. a straight-line piece of code without any statements dealing with conditions or a conditional statement, and* A *is a set of directed arcs. Vertices* v_1 *and* v_2 *are said to be connected if there is a direct control flow in the program between the corresponding statements for* v_1 *and* v_2. *Moreover, we assume the designated vertices ENTRY and EXIT to be element of* V . *ENTRY represents the beginning of the control flow and EXIT the end of it. If a vertex* $x \in V$ *is a conditional, then all arcs* $(x, y) \in A$ *are labeled with* true *or* false *which corresponds to control flow of the program.*

The CFG of the *runM2M()* procedure is depicted in Fig. 6.2. Formally, the problem of converting programs into rules that express the possible behavior can be stated as follows.

Definition 2 (Conversion problem) *The conversion problem is specified by* (Π, F) *where* Π *is a control program (written in C) and* F *is a set of procedures or functions of interest.*

A solution to the conversion problem is a set of rules which represents possible execution sequences of the program in order to execute the functions specified in F. We assume that a control program basically comprises an infinite loop where the procedures given in F are executed depending on certain conditions. Hence, only the loop-free part of the program has to be considered. In cases where procedures of interest are called in a procedure call of the body of the infinite loop it is easy to compile such a program into one where this is not the case. This might be done by replacing the procedure call with the body of the procedure.

Since we are interested in the executions of the procedures or functions in F, we have to get the information regarding conditions under which such a function is guaranteed to be executed. When having a look at the CFG in Fig. 6.2, we see that this information corresponds to the possible paths from the ENTRY vertex to the vertex where a function call to a $f \in F$ is given in the corresponding source code.

Definition 3 (Solution) *Given a conversion problem* (Π, F). *A solution of* (Π, F) *is a set of rules of the form* $c_1 \wedge \ldots \wedge c_n \rightarrow f$ *where* $c_1, \ldots, c_n$ *are logical literals (conditional expressions used in the program or their negation) which have to be valid in order to execute* $f \in F$.

As already discussed we want conditionals used in the rule not to reference local variables. Hence, we restrict solutions to rules comprising conditional expressions not using local variables.

Definition 4 (Restricted solutions) *A solution of a conversion problem* (Π, F) *is said to be restricted if no conditional expression of any rule references a local variable.*

The reason for this restriction is that we want to replace the control program by a set of rules where each part of the rule corresponds to a function that can be executed. The following algorithm computes restricted solutions.

6.3.1 Algorithm ComputeRules

Input: A program Π and a set of procedures or functions of interest F.
Output: A set of rules.

(1) Let Π' be the program where all local variables used in conditional expression of Π have been eliminated by using behavior preserving transformations.
(2) Construct a CFG for Π'.
(3) Let R be the empty set. In R we are storing the extracted rules.
(4) For all $f \in F$ do:
  (a) For all vertices v where f is called in the corresponding source code do:
    (i) Extract the path(s) $(ENTRY, v_1, \ldots, v_k, v)$ from ENTRY to the vertex v.
    (ii) Apply the transformation function l to the path(s) which is defined as follows:

$$l(x) = \begin{cases} \in & \text{if x = v or x = ENTRY or x is not a conditional} \\ x & \text{if y is the immediate successor of x in the path and the label of the arc (x, y) is true} \\ \neg x & \text{if y is the immediate successor of x in the path and the label of the arc (x, y) is false} \end{cases}$$

  Let $(l_1, \ldots, l_k)$ be the path after applying the function l.

    (iii) Generate rule(s) $l_1 \wedge \ldots \wedge l_k \rightarrow f$ and add it to the set of rules R.
(5) Minimize the set of rules R and return them as result.

It is obvious that **computeRules** terminates because the CFG is finite and contains no loops, there are only a finite number of paths and the set F is also finite. Because we are searching for all paths **computeRules** is in the worst case exponential in the number of conditions. Further improvements which should consider the structure of the CFG are left for future research. Note that for step 5 standard rules of propositional calculus, such as Modus Ponens, can be used. In the rest of this section we focus on step 1 of algorithm **computeRules**.

6.3.2 Program Transformation

For the rule computation part of **computeRules** it is necessary that all conditions must not use local variables in order to ensure these programs can be transformed. Transforming a program by hand is always possible because this only requires introducing global variables instead of local ones. In cases of name conflicts new names have to be introduced. However, we are more interested in an automatic transformation which is based on the idea of encapsulating program parts that form a semantic entity. In our running example, lines 18–23 form such an entity. All statements of this part contribute to the detection of a position change using GSM data. But how can this transformation be done automatically? We do not give a general answer in this article but we want to state a case where this behavior preserving transformation can be obtained from static analysis.

The idea behind this transformation is as follows. Consider the CFG of our running example. Here statements 18–22 are represented by a vertex v of the CFG and line 23 by another vertex v' which is the immediate successor of v. When analyzing the data dependencies in the program, e.g., using methods from slicing [2,3], we know that the local variables referenced in line 23 (and vertex v') are neither defined nor used in another part of the program. In this case the situation is much easier because of the fact that the scope of the variables *new* and *old* is restricted to the sub-block. Hence, we are able to generate a new function which comprises the statements of vertices v and v' and replace them by a respective function call.

In general transformations can be represented by a rule of the form $\langle pre-condition \rangle : \langle transformation \rangle$. Whenever the pre-condition is fulfilled the transformation can be applied. In our specialized case we have the following transformation rule:

Pre-condition Given the vertices v and v' where v' is the immediate successor of v. v' has to be a conditional vertex where the corresponding referenced variables are defined in v and are referenced only in v or v'.

Transformation

(a) Use static analysis to determine variables V that are not global, defined neither in v or v' but used in either v or v'. For all $x \in V$ generate a parameter list AS separated by ',' where every element is of the form '$type(x)\ x$'. In addition generate a similar string VS for stating all parameters of the function call where each element is of the form 'x'.
(b) Generate the body of the function. Add the source code of v to B. Add the condition 'if C return true else return false' to B where C is the corresponding condition of v'.
(c) Introduce an unused procedure name id and generate the function definition *int* id (AS) { B }.
(d) Remove the code corresponding to v from the original program. Replace the conditions of v' with the condition 'id (VS)'.

This concludes the discussion of the conversion algorithm. In the subsequent sections, we focus on experiences gained when working with an implementation of the presented approach.

6.4 Case Study and Discussion

We have implemented the presented approach and used it for extracting rules from the control program we already presented excerpts from. The tool expects a configuration file that specifies the source files to look at, a set of atomic functions, a set of target functions, and the name of a function that serves as starting point for the extraction process.

Until now, we have assumed to start our extraction process from the *main* function only. Often times, however, control programs have more than one entry point, as interrupt service routines usually serve as program entry point too. Because our tool can be configured to use an arbitrary function as starting point, we are able to look at interrupt service routines for the purpose of knowledge extraction.

The set of target functions the tool expects corresponds to the *functions of interest* as defined earlier. Internally, the tool treats the set merely as a sequence, doing rule extraction for one function of interest at a time.

As building up the control flow graph and extracting all paths of how to reach a certain function of interest may get quite costly, it is possible to specify a set of atomic functions. Specifying a function as atomic has the effect that the tool does not try to look inside the function but only includes the calls of the function in the control flow graph. Effectively this optimization leads to shorter (in terms of included statements) paths when traversing the control flow graph for rule extraction. Of course, by declaring a function atomic, we may also miss paths to the function of interest, which later on means that the function of interest may also be called from low-level code.

After processing the source files, the tool outputs a list of conditions that need to be fulfilled in order to reach the function of interest. It also provides the necessary C-code for these conditions. Taking our running example and extracting rules for *makeActiveCall* leads to following result:

```
makeActiveCall <- do1, do2, do5, check6, do7, !check8

makeActiveCall <- do1, do2, do5, check6, do7, check8,
do9, !check10
```

In the above rules, the comma denotes a logical and ($\wedge$), *doX* stands for some action, i.e. C-code, that is executed after all conditions evaluate to true and the rule is "run", *checkX* for conditions that have to be checked before a rule can be selected and "run", and "*!*" represents logical negation ($\neg$). Ignoring the actions for the moment, and joining the two rules into one, we derive

$$makeActiveCall \leftarrow (check6 \wedge \neg check8) \vee (check6 \wedge check8 \wedge \neg check10)$$

Simplifying the expression yields

$$makeActiveCall \leftarrow check6 \wedge (\neg check8 \vee \neg check10)$$

which is equivalent to the manually calculated result when substituting $cond_2$ for *check6*, $cond_3$ for *check8*, and $cond_4$ for *check10*. Next we show that this substitution is valid by looking at the extracted C-code.

When looking at the code generated for *do1*, we notice that this is initialization code taken from the main function. Because we're not interested in one-time initialization code, *do1* can be ignored. The same argument holds for *do2*. The next action, *do5*, is different:

```
int do5() {
   if (getGSMEvent ( GSM_EVENT_RING ) ) {
      makePassiveCall ( ) ;
      m2mReportTimer = REPORT_TIMEOUT ; }
   return (1 == 1); }
```

This action hides the call to *makePassiveCall*: Because we've only looked for rules regarding *makeActiveCall* and the behavior of the generated rule-set should closely match the behavior of the original C code, this action is included. Of course, when also extracting rules for *makePassiveCall*, *do5* can be ignored. Next, the algorithm has extracted a first condition, namely *check6*. This condition returns *return (! getM2MReportTimer())*, which is what we expect. The action *do7* is only meaningful when the rule is run and sets a new value to the *m2mReportTimer*. *Check8* returns the result of *getGPSState( ) == 'S')*, while *do9* is again some internal action. Most interesting is *check10*, as the condition is derived from

```
if (old == new) return;
```

Instead of *old == new*, the tool returns a condition *check10* defined as

```
int __checkCondition10__() {
   char old , new ;
   old = getSignalQuality ( ) ;
   sendGSMCommand_CSQ ( ) ;
   waitM2MSeconds ( 3 ) ;
   new = getSignalQuality ( ) ;
   return (old == new ); }
```

which is exactly what we need. From the given example it is obvious that the output of the tool still needs to be checked manually. For example, the extracted conditions and actions lack meaningful names which force the developer to revise the extracted rule set. That said the tool helps the developer to get a quick understanding of the code – in particular how complex the rules will be, when no simplification is performed.

Besides these technical issues, there are some more fundamental limitations that have to be kept in mind when working with our tool: Given an atomic function c that sets global variables i and j to $i = j$. Given further following program fragment:

```
if (i != j) {
   if(a) c;
   if (i==j) return;
   if (!a) c;
}
```

Due to the side effect, line four is never looked at when running the program and '*a*' is true. In this case, line four is dead code. Our algorithm, using static analysis only, will extract following rules: $R_1 : i <> j \wedge a \rightarrow c$

and $R_2: i <> j \wedge !a \rightarrow c$ which simplifies into following minimal rule set: $i <> j \rightarrow c$. Clearly, this is not the intended behavior. One might think that a non-minimal rule set indicates that the program has been incorrectly converted and the behavior of the rule based system does not match the one generated by the C-Code version. This is, of course, not the case as a correct (with respect to some specification) C program can be given that does the same operation in both branches of an "if" statement. Thus, the extracted rule set will not be minimal. Minimality of the extracted rule base, however, does not guarantee correctness either. This quickly can be seen by exchanging the last call to c in the presented program fragment by some call of d. The extracted rule set will now be minimal, but won't match the behavior of the C program. Other difficulties that require manual review are the usual suspects, such as interrupt handling, loops, and the fact that by extracting rules we've applied abstraction to the control program and therefore might not replicate the original behavior, also demonstrated in the previous example.

6.5 Related Research

To the best of our knowledge, however, none of these techniques aim for a replacement of high-level program parts while, at the same time, linking with unchanged low-level parts of the same program. Tools like BLAST [4], SLAM [5] and CMBC [6] are used to check safety properties in software and return a set of conditions under which the safety property is hurt. In difference to our algorithm, these tools only supply one trace of how to reach a certain function while we require all paths to be known.

Tools like DAIKON [7], or DySy [8] can detect program invariants but need to run the program in order to analyze it. Besides the nontrivial issue of running and analyzing embedded firmware within a simulator, the detected program invariants require manual check. Even if these issues are solved, we still need to transform the original C-program to work with the findings.

C-Wolf [9], for example extracts an abstract labeled transition system of C-code. In a sense this work is similar, because our rule set also specifies possible transitions in the system. That said we do not want to replace the original control program but only high-level parts, which is more related to specification mining [10], which, in turn, has its own requirements.

Finally, the L* algorithm, introduced by Angluin [11], creates a minimal DFA for an unknown regular language it has learned. The structure of the DFA is inferred by asking a teacher who knows the language membership and equivalence queries. As our working hypothesis is that the original control program implements a DFA, it seems reasonable to use L* and the developer in order to extract rules. While this seems promising, we still have to problem of code reuse.

6.6 Conclusion

In this chapter we present results regarding the extraction of rule-based knowledge directly from control programs written in C. The underlying idea relies on the special structural properties of control programs which basically comprise a loop statement and conditional statements. The extraction process takes as input a set of function or procedures which are assumed to capture the required high-level knowledge. The main advantage of the approach is the extraction of high-level rules (when do what) that can seamlessly interact with original low-level C-Code. This guarantees a maximum amount of code re-use when switching to a smart runtime driven hardware design and is the main motivation for the presented approach.

References

1. W. Krenn and F. Wotawa, Gradient-based diagnosis, In Proceedings of the International Workshop on Principles of Diagnosis, pages 314–321, 2007.
2. M. Weiser, Programmers Use Slices when Debugging, Communications of the ACM, 25(7):446–452, 1982.
3. M. Weiser, Program Slicing, IEEE Transactions on Software Engineering, 10(4):352–357, 1984.
4. T. Henzinger, R. Jhala, R. Majumdar, and G. Sutre, Software verification with Blast, Proceedings of the Tenth International Workshop on Model Checking of Software (SPIN), Lecture Notes in Computer Science 2648, Springer-Verlag 2003.
5. T. Ball and S. K. Rajamani, The SLAM project: debugging system software via static analysis, Proceedings of the 29th ACM SIGPLAN-SIGACT symposium on Principles of programming languages, pages 1–3, Jan. 2002.
6. C. Edmund, D. Kroening, and F. Lerda, Flavio, A Tool for Checking ANSI-C Programs, Tools and Algorithms for the Construction and Analysis of System, 2004.
7. M. D. Ernst, J. H. Perkins, P. J. Guo, S. McCamant, C. Pacheco, M. S. Tschantz, and C. Xiao, The Daikon system for dynamic detection of likely invariants, Science of Computer Programming, 2007.
8. C. Csallner, N. Tillmann, and Y. Smaragdakis. Dysy: Dynamic Symbolic Execution for Invariant Inference. In Robby, editor, ICSE, pages 281–290, ACM, 2008.
9. D. Du Varney and S. Purushothaman Iyer, C Wolf – A Toolset for Extracting Models from C Programs, Formal Techniques for Networked and Distributed Sytems, 2002.
10. S. Shoham, E. Yahav, S. Fink, and M. Pistoia, Static Specification Mining Using Automata-Based Abstractions, ISSTA, 2007.
11. D. Angluin, Learning regular sets from queries and counterexamples, Information and Computation, 75(2):87–106, 1987.

Chapter 7

Real Time Implementation of Fuzz-Face Electric Guitar Effect

Massimo Conti[1], Simone Orcioni[1], Marco Caldari[2] and Franco Ripa[2]
[1]*D.I.B.E.T., Università Politecnica delle Marche, Ancona, Italy, m.conti@univpm.it*
[2]*Korg Italy, Osimo, Italy*

Abstract Physical models are widely used for sound synthesis and transformation. This chapter presents a general methodology for the integration of physical modeling of sounds in a system level design environment using SystemC. The methodology has been applied in particular for physical modeling of electric guitar effects, derived from the well known fuzz-face circuit. The implementation in real time systems of physical models requires very high performance processors and dedicated hardware. The implementation on low cost embedded systems implies a further simplification of the algorithms. This paper presents the implementation of electric guitar effects in an embedded system board based on the ARM7 processor.

Keywords Physical models, SystemC, Electric guitar effects, Fuzz-face, Embedded system

7.1 Introduction

Many techniques used nowadays for the electronic generation and processing of sounds trays to generate directly the output waves, for example the Frequency Modulation, additive or subtractive synthesis, wavetable synthesis based on the storage of samples of sounds of real musical instruments [1–5]. In spite of the simplicity of their implementation, the sound quality cannot meet the requirements of the most demanding users.

An alternative technique, called physical modeling, tries to identify the physical mechanism that generates sounds, starting from the physical laws that are on the basis of the instrument. In the field of musical instruments

N. Martínez Madrid, R.E.D. Seepold (eds.), *Intelligent Technical Systems*, Lecture Notes in Electrical Engineering 38, DOI 978-1-4020-9823-9_7,

this technique is used to define algorithms for the synthesis or transformation of sounds.

Physical modeling allows good sounds and gives a natural and intuitive interaction between the player and sound generator since the model focuses on the sound production mechanism rather than on the sound itself. A physical model starts from the mechanical or fluidodynamical reference model that is represented by differential equations.

The most commonly used physical modeling technique is called Wave Digital Filter (WDF) [6–9]. The WDF algorithm starts from the wave equation of musical strings and implements the solution of the equation on a discrete-time system.

Many applications of the WDF technique have been developed, for example for the synthesis of a cello [10], hammer string interaction in an acoustic piano [11], string [12], human singing [13].

Physical models sometimes lead to computational expensive algorithms. With the increasing computation capabilities of DSPs, this is no more a problem for real time applications. Recent works present the implementation of WDF on FPGA in order to accelerate the numerical methods to allow real-time production of sounds for musical applications [14,15].

Actually a unique design framework allowing the simulation of the mechanical physical model, the simulation of the simplified equivalent electric physical model, the design and simulation of the digital discrete time algorithm and the design, simulation and optimization of the implementation on DSP and/or FPGA is not available.

This design environment requires the cosimulation of systems of different nature: mechanical, analog electronic, digital discrete time electronic systems. SystemC is a consolidated design language and environment based on C++ used for a system level description of System on Chip [16]. The extension of SystemC to mixed-signal is under development. The aim of this extension is the codesign and cosimulation of mechanical and electronic analog and electronic digital parts. Application examples are systems in which the analog RF part is integrated in the same chip with the digital part, or a digital control of the injection system in the automotive field. The SystemC-AMS working group inside OSCI is working in the development of a new library to be integrated in SystemC for the description of mixed-signal systems [17–20].

Recently the SystemC-WMS (Wave Mixed Signal) library has been presented [21,22], allowing a simple SystemC extension to mixed-signal using the concept of incident and reflected waves. SystemC-WMS allows a simple simulation environment of WDF for sound synthesis. The SystemC-WMS model [23] of an electric guitar effect has been presented.

This chapter presents a SystemC description of the well known fuzz-face electric guitar distortion effect. The algorithm has been simplified in order to allow the implementation in an embedded system on a board based on the ARM7 processor.

Section 7.2 describes the fuzz-face circuit implementing the electric guitar effect. Section 7.3 shows the design methodology used. In Section 7.4 an accurate SystemC model is reported, while a simplified implementation for the ARM7 is reported in Section 7.5.

7.2 The Fuzz-Face Electric Guitar Effect

This work presents a distortion effect for electric guitar derived from the analog distortion circuit, reported in Fig. 7.1, called Fuzz-Face [24], produced by Dallas Arbiter in 1966 and famous in the 1960 and 1970 and used by Jimi Hendrix, David Gilmour, Carlos Santana, Eric Clapton, and widely used up to now.

The circuit saturates for some values of the input signal, giving the effect of distorting a sinusoid into an almost square waveform introducing new harmonic components.

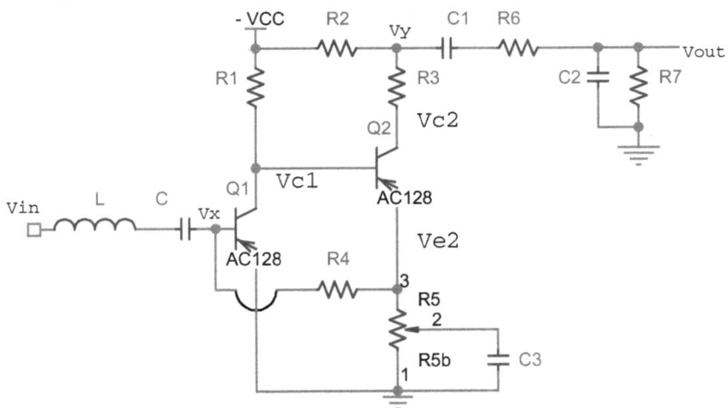

Fig. 7.1 Fuzz-Face circuit for electric guitar effect

The original and the actual analog circuital implementations use two coupled germanium transistors, which are preferred due to their better "sound performances" to the cheaper silicon transistors. The germanium transistors suffer from strong dependence on temperature and humidity, high variability of the parameters, high probability of breaking during soldering, high cost due to the fact that germanium is an obsolete technology. The advantage of a discrete time digital implementation could be the cost reduction, easy tuning of the parameters by the player himself due the physical model nature of the algorithm, easy integration in a digital sound generation and processing.

7.3 Design Methodology

The design flow applied in this work is reported in Fig. 7.2; it is applied for a sound transformation system, but it can be applied for sound generation system as well.

The design starts from the analysis of the real analog fuzz-face circuit. In a second step, a spice model of the electric guitar effects generator is derived, and the coefficients of the model have been tuned in order to fit the experimental data.

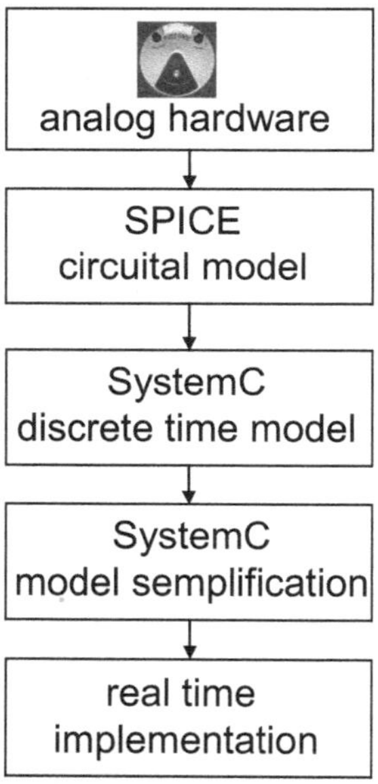

Fig. 7.2 Design flow for physical modeling

The next step consists in the creation of a discrete time Signal Flow Graph model in SystemC. Further model simplification and architecture refinement must be performed in order to allow the implementation on an embedded system, in this case the ARM7 microprocessor chosen for the real time implementation.

The final step is the definition of the resolution of the variables, and the definition of the architecture. The translation from the SystemC description in the C code for the ARM7 is very easy.

The design methodology allows a simple system level reuse and integration with existing hardware or other sound processing or generation algorithms already developed.

The model developed at different levels of abstraction can be tested in real time.

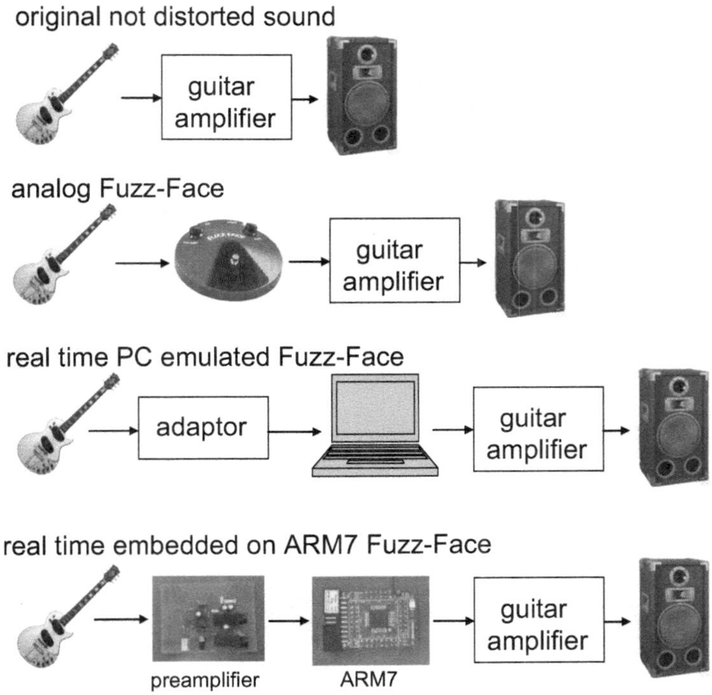

Fig. 7.3 Real time implementations of the fuzz-face distortion

Figure 7.3 shows the different real time implementations developed:

- the original sound of the electric guitar.
- the analog fuzz-face circuit.
- the SystemC discrete time model applied in real time to the electric guitar through an input and output adaptor to the PC. A real time implementation is possible if the complexity of the model is not high compared to the computational performances of the hosting PC, as in the case presented in this work.
- the embedded digital system based on an ARM7 processor.

A real time test is fundamental to speed up the optimization and to verify the quality of the sound in a real environment.

7.4 Systemc Discrete Time Model

The simplified model implemented in SystemC using a discrete time model consists in three modules, as reported in Fig. 7.4. The main simplifications with respect to the original circuit reported Fig. 7.1 are:

- the current absorbed by the output modules have been neglected, in this way the system can be represented in by a signal flow graph, enabling a simple representation using SystemC modules;
- discretization of the differential equation using the forward Eulero formulae;
- solution of the nonlinear equations of the Fuzz-Face module using the Newton-Rapson iteration scheme.

The differential equations describing the pickup module, in Fig. 7.4, are the following

$$v_{in} - v_x - v_c = \frac{L}{R}\frac{dv_x}{dt} \qquad (7.1)$$
$$v_x = RC\frac{dv_c}{dt}$$

A constant resistance R has been used to model the nonlinear load of the fuzz-face of Fig. 7.1. The differential equation representing the output filter module is the following

$$\frac{v_z - v_{out}}{R_6} - \frac{v_{out}}{R_7} = C_2\frac{dv_{out}}{dt} \qquad (7.2)$$
$$\frac{v_z - v_{out}}{R_6} = C_1\frac{d(v_y - v_z)}{dt}$$

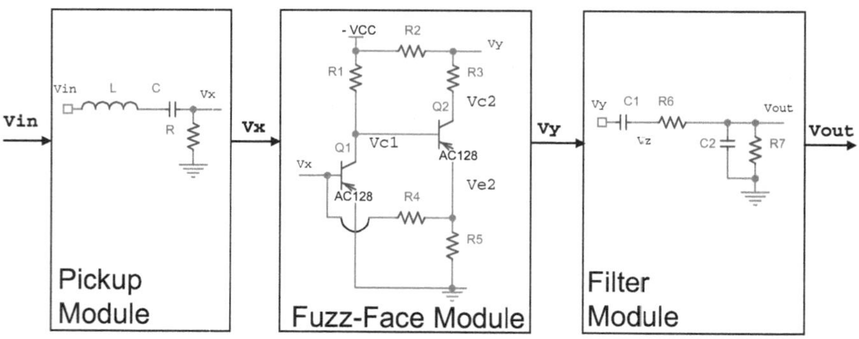

Fig. 7.4 SystemC discrete time signal flow graph model of the fuzz-face

The differential equations (7.1) and (7.2) have been discretized using the Eulero formula as indicated in equation (7.3) and considering as time step h the sampling time of 22.6 μs, corresponding to the 44.1 kHz sampling rate normally used for audio signals.

$$x = \frac{dy}{dt} \quad \Rightarrow \quad \frac{x(t)+x(t-h)}{2} = \frac{y(t)-y(t-h)}{h} \tag{7.3}$$

The Fuzz-Face distortion module is described by the nonlinear equations reported in equations (7.4), obtained using the Kirchoff laws and considering the variables v_{c1} (the collector voltage of Q1) v_{c2} and v_{e2} (the collector and emitter voltages of Q2).

$$\begin{aligned}
&\alpha_{F1} I_{E01}\left(\exp\left(\frac{-v_{in}}{v_t}\right)-1\right) - I_{C01}\left(\exp\left(\frac{v_{c1}-v_{in}}{v_t}\right)-1\right) + I_{E02}\left(\exp\left(\frac{v_{e2}-v_{c1}}{v_t}\right)-1\right)(1-\alpha_{F2}) \\
&\qquad + I_{C02}\left(\exp\left(\frac{v_{c2}-v_{c1}}{v_t}\right)-1\right)(1-\alpha_{R2}) - \frac{v_{c1}}{R1} = 0 \\
&I_{E02}\left(\exp\left(\frac{v_{e2}-v_{c1}}{v_t}\right)-1\right) - \alpha_{R2} I_{C02}\left(\exp\left(\frac{v_{c2}-v_{c1}}{v_t}\right)-1\right) - \frac{(v_{in}-v_{e2})}{R4} + \frac{v_{e2}}{R5} = 0 \\
&\alpha_{F2} I_{E02}\left(\exp\left(\frac{v_{c2}-v_{c1}}{v_t}\right)-1\right) - \frac{(V_{CC}+v_{c2})}{R2+R3} = 0
\end{aligned} \tag{7.4}$$

The Newton-Rapson iteration scheme has been used to solve equations (7.4). The equations (7.4) can be written as follows

$$\begin{aligned}
f_1(x_1,x_2,x_3) &= 0 \\
f_2(x_1,x_2,x_3) &= 0 \\
f_3(x_1,x_2,x_3) &= 0
\end{aligned} \tag{7.5}$$

or in a more compact way

$$F(X) = 0 \tag{7.6}$$

Where $X = (x_1, x_2, x_3)^T = (v_{c1}, v_{c2}, v_{e2})^T$ (7.7)

The Newton-Rapson algorithm can be written as follows

$$X^{(n)} = X^{(n-1)} - J^{-1}(X^{(n-1)})F(X^{(n-1)}) \tag{7.8}$$

where J is the jacobian matrix

$$J = \begin{bmatrix} \frac{\partial f_1}{\partial x_1} & \frac{\partial f_1}{\partial x_2} & \frac{\partial f_1}{\partial x_3} \\ \frac{\partial f_2}{\partial x_1} & \frac{\partial f_2}{\partial x_2} & \frac{\partial f_2}{\partial x_3} \\ \frac{\partial f_3}{\partial x_1} & \frac{\partial f_3}{\partial x_2} & \frac{\partial f_3}{\partial x_3} \end{bmatrix} \tag{7.9}$$

Equation (7.8) has been implemented in the SystemC module. Once v_{c2} is obtained, v_{out} is derived using equation (7.10)

$$v_{out} = v_{c2}\frac{R2}{R2+R3} - V_{CC}\frac{R3}{R2+R3} \tag{7.10}$$

The discrete time SystemC model has been integrated with other libraries allowing audio real time simulations.

The input waveform can be taken from a file or directly from the guitar connected through an adapter to the PC, and the output can be stored into a file or directed to audio output of the PC, as shown in Fig. 7.3. Physical modeling allowed the definition of parameters of the model that are directly related to the physical mechanism that generated the distortion. These parameters can be changed in real time while the artist is playing his guitar.

This allows real time verification by the sound experts of the quality of the sound effects, giving a feedback to the electronic designer from the first step of the design refinement. The CPU usage of the portable PC used is about 30%.

The response of the circuit to an input sinusoid from Spice simulations and SystemC simulations with a discrete time description are reported in Fig. 7.5. The comparison between the simulations allows us to tune the parameters of the SystemC model and verify the agreement of the discrete time physical model with the real circuit.

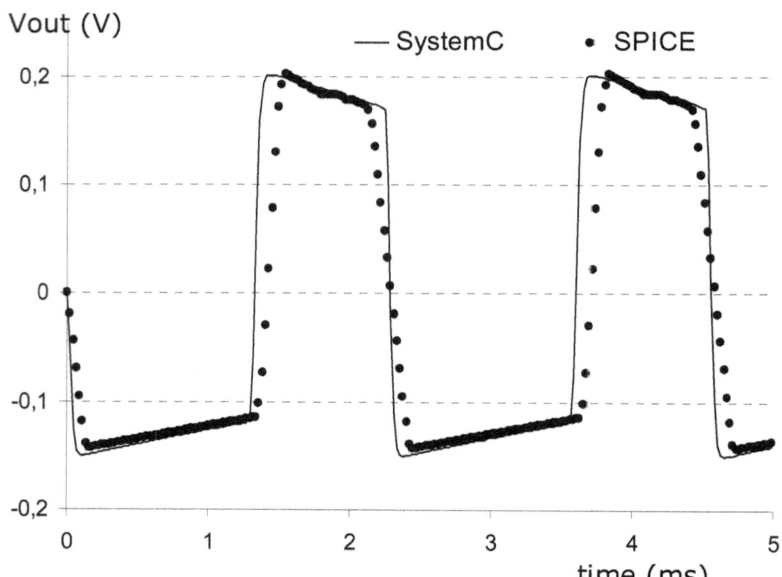

Fig. 7.5 SPICE and SystemC (using a discrete time description) simulations of the Fuzz-Face circuit with a sinusoidal input

The real time simulation environment has been tested with a real electric guitar. Figure 7.6 shows the waveforms of the electric guitar without distortion (original), the signal after the distortion of the Fuzz-Face implemented in hardware with germanium transistors (real Fuzz-Face) and the signal after the distortion obtained in real time with software developed (virtual Fuzz-Face).

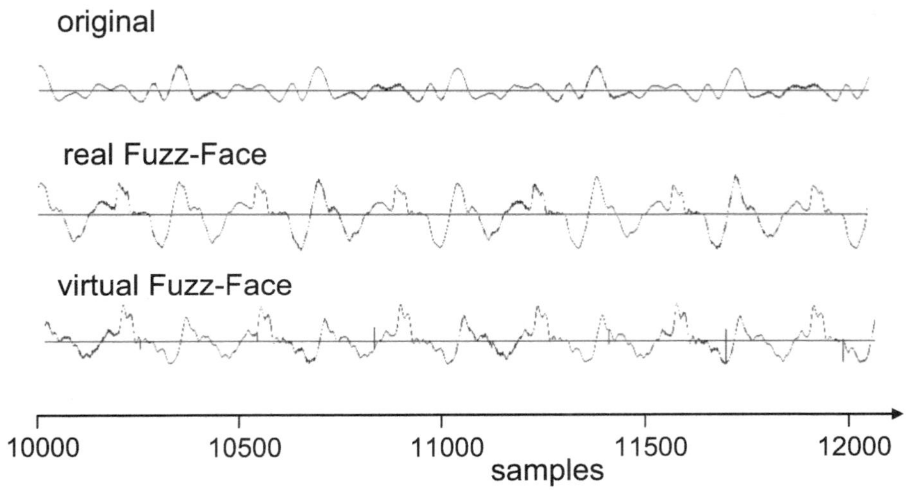

Fig. 7.6 Waveforms of the electric guitar without distortion (original), after the distortion of the hardware Fuzz-Face (real Fuzz-Face) and the distortion obtained with software developed (virtual Fuzz-Face)

7.5 Implementation on Embedded Systems

The embedded system chosen to implement the fuzz-face is the microcontroller LPC2148 of NXP, based on a 16-bit/32-bit ARM7TDMI-S CPU 60 MHz maximum CPU clock, shown in Fig. 7.7. Various 32-bit timers, one 10-bit DAC, PWM channels are provided. Two 10-bit ADCs provide a total of 14 analog inputs, with conversion times up to 2.44 us per channel.

In spite of the simplification of the discrete time SystemC model the real time implementation on an embedded system is not possible.

Equations (4.10) require the calculus of exponential functions, multiplications and divisions that are complex functions to be implemented.

The first solution we tried is the implementation equations (4.10) using a 32 bit fixed point representation. The fixed point solution gives accurate solution but the CPU time required for an implementation on a portable PC with a Pentium M processor is still too expensive: for example the processing of a sound 12 s long takes about 9 s. Therefore the real time implementation of the floating point or fixed point algorithm on an embedded ARM7 processor with a 60MHz clock frequency is not possible.

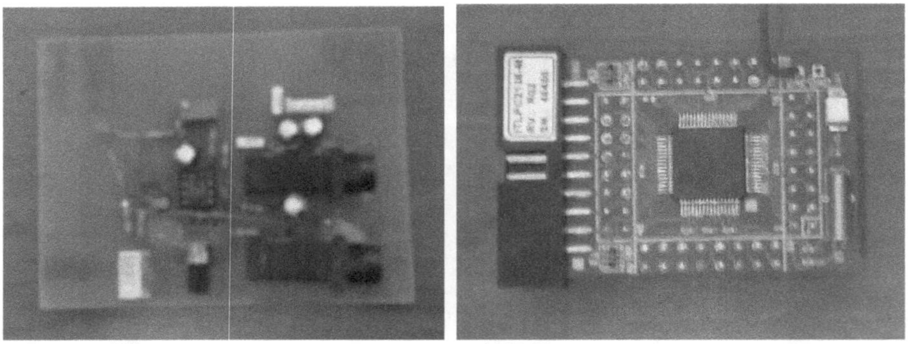

Preamplifier ARM7

Fig. 7.7 The electronic board and its main modules

The distortion algorithm represented in Fig. 7.4 has been simplified approximating the nonlinear static relationship of the complete fuzz-face block with a look up table, as shown in f Fig. 7.8. The pickup and filter modules are the same as the ones in Fig. 7.4. A 32 bit fixed point representation has been used. The ARM7 performs an elaboration on each input sample with a sampling time of 22.6 μs (44.1 kHz), the calculus requires 5.33 μs, therefore a real time implementation is possible.

The same elaboration on the ARM7 using a floating point representation requires about 32 μs, that not allows a real time implementation.

The sinusoidal input and the distorted output of the embedded system board are reported in Fig. 7.9.

7.6 Conclusions

This paper presents a design flow for the design of sound digital processing circuits based on physical models. The real time simulation of the algorithmic description allows a simple verification of the quality of the sound effect by the designer and by the guitar player. Following the presented methodology, the digital model of the Fuzz-Face has been implemented in real time on the embedded system LPC2148 of NXP, based on a 16-bit/32-bit ARM7. Future work will be devoted in the refinement of the algorithm to improve the quality of the sound.

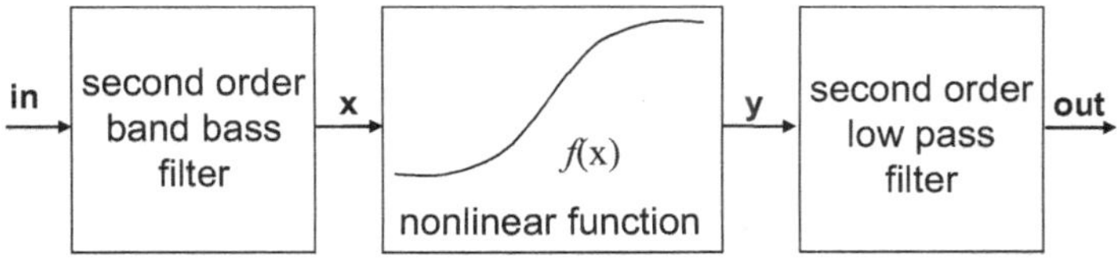

Fig. 7.8 SystemC simplified model

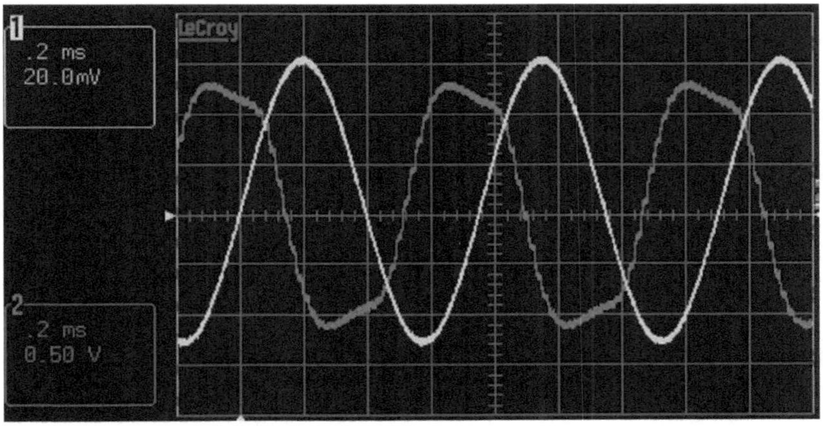

Fig. 7.9 Sinusoidal input and distorted output of the ARM7 real time implemenation

References

1. B. L. Vercoe, W. G. Gardner and E. D. Scheirer, "Structured audio: Creation, transmission, and rendering of parametric sound representations", Proceedings of the IEEE, VOL. 86, NO. 5, MAY 1998, pp. 922–940.
2. H.G. Alles, "Music sysnthesis using real time digital techniques", Proc. Of IEEE, Vol.68, No. 4, april 1980, pp. 436–449.
3. E. Holsinger, "How Music and Computers Work", Chicago, IL: Ziff- Davis Press, 1994.
4. S. Pellman, "An Introduction to the Creation of Electroacoustic Music", Belmont, CA: Wadsworth, 1994.
5. J.M. Chowning, "The synthesis of complex audio spectra by means of frequency modulation," J. Audio Eng. Soc., vol. 21, no. 7, pp. 526–534,1973.
6. J.O. Smith, "Physical Modeling using Digital Waveguides", Comput. Music J., special issue on Physical Modeling of Musical Instruments, Part I, Volume 16, No. 4, p. 74, 1992.
7. J.O. Smith, "Music application of digital waveguide," Stanford Univ., CCRMA Tech. Rep. STAN-M-67.
8. S. A. Van Duyne and J. O. Smith, "Physical modeling with the 2-D digital wave guide mesh," in Proc. Int. Computer Music Conf., Tokyo, Japan, 1993, pp. 40–47.

9. S. Petrausch, J. Escolano and R. Rabenstein, "A general approach to block"-based physical modeling with mixed modeling strategies for digital sound synthesis", Proc. of ICASSP '05, Volume 3, 18–23 March 2005, pp. 21–24, Vol 3.
10. Sheng-Fu Liang, Alvin W. Y. Su and Chin-Teng Lin, "Model-based synthesis of plucked string instruments by using a class of scattering recurrent networks", IEEE Trans. On Neural Networks, Vol. 11, No. 1, Jan 2000, pp. 171–185.
11. Pedersini, F., Sarti, A. and Tubaro, S., "Block-wise physical model synthesis for musical acoustics", Electr. Lett., Vol. 35, No. 17, 19 Aug. 1999 Page(s):1418–9.
12. J. Escolano and J.-J. Lopez, "On the adaptation of the linear bicharacteristic scheme to block-based physical modeling for digital sound synthesis of string instruments", Proc of ICASSP 2006, pp.V-161–164.
13. C. Cooper, D. Murphy, D. Howard and A. Tyrrell, "Singing synthesis with an evolved physical model", IEEE Trans. on Audio, Speech and Language Processing, Vol. 14, No. 4, July 2006, pp. 1454–1461.
14. E. Motuk, R. Woods and S. Bilbao, "FPGA-based hardware for physical modeling sound synthesis by finite difference schemes", Proceedings. 2005 IEEE International Conference on Field-Programmable Technology, 2005, 11–14 Dec. 2005, pp. 103–110.
15. J.A. Gibbons, D.M. Howard and A.M. Tyrrell, "FPGA implementation of 1D wave equation for real-time audio synthesis", IEEE Proc. of Computers and Digital Techniques, Volume 152, Issue 5, 9 Sept. 2005, pp. 619–631.
16. SystemC, www.systemc.org, 2008.
17. A. Vachoux, C. Grimm and K. Einwich, "SystemC-AMS requirements, design objectives and rationale", Design, Automation and Test in Europe Conference, 2003, pp. 388–393.
18. A. Vachoux, C. Grimm and K. Einwich, "Towards analog and mixed-signal SOC design with systemC-AMS", Workshop on Electronic Design, Test and Applications, 2004. DELTA 2004, pp. 97–102.
19. OSCI Analog/Mixed-signal Working Group (AMSWG), www.systemc.org/projects/ams-wg/
20. SystemC-AMS, http://www.systemc-ams.org/, 2008.
21. S. Orcioni, G. Biagetti and M. Conti, "SystemC-WMS: Mixed Signal Simulation based on Wave exchanges", in the book "Advances in design and specification languages for SOCS", Alain Vachoux (Editor.), Kluwer Academic 2006.
22. M. Conti, M. Caldari, S. Orcioni and G. Biagetti, "Analog circuit modeling in SystemC", in the book "Languages for System Specification and Verification" CHDL Series, Christoph Grimm (Editor.), Kluwer Academic 2004, pp. 229–242.
23. F. Gambini, M. Conti, S. Orcioni, F. Ripa and M. Caldari, "Physical modelling in SystemC-WMS and real time synthesis of electric guitar effects", Proc. of the WISES07, pp. 87–100, Madrid, Spain, June 21–22, 2007.
24. R.G. Keen. "The technology of the Fuzz-Face", www.geofex.com/Article_Folders/fuzzface/fffram.htm, 2008.

Chapter 8

Providing Standardized Fixed-Point Arithmetics for Embedded C Programs

Wilfried Elmenreich[1], Andreas Wolf[2] and Maximilian Rosenblattl[2]
[1]*Lakeside Labs, Mobile Systems Group, Institute of Networked and Embedded Systems, University of Klagenfurt, 9020 Klagenfurt, Austria, wilfried.elmenreich@uni-klu.ac.at*
[2]*Vienna University of Technology, 1040 Vienna, Austria*

Abstract The ISO/IEC Standard TR 18037 defines the syntax and semantics for fixed-point operations for programming embedded hardware in C. However, there are currently only few compilers available that support this standard. Therefore, we have implemented a stand-alone library according to the standard that can be compiled with standard C compilers. The library is available as open source and written in plain C, thus can be used in various target architectures as long as a C compiler is available. This book chapter presents a brief description of the ISO/IEC standard and the library implementation followed by an evaluation of code size and performance of the fixed-point operations on the Atmel AVR architecture. A comparison with the standard floating-point library (which is machine code-optimized to the target architecture) shows that simple fixed-point functions such as addition, subtraction and multiplication are more efficient, while more complicate functions can only compete in the worst case behavior. The fixed-point approach provides a smaller memory foot print, for typical applications where only a small subset of functions is used. This is especially of interest for the big market of embedded microcontrollers with only a few Kbytes of program memory.

Keywords Fixed-point arithmetic, C Programming language, Embedded C, CORDIC

8.1 Introduction

The C language standard [1] specifies two data types for expressing fractional numbers, the float and the double data type. Both data types are in

N. Martínez Madrid, R.E.D. Seepold (eds.), *Intelligent Technical Systems*, Lecture Notes in Electrical Engineering 38, DOI 978-1-4020-9823-9_8,

a floating-point format consisting of sign, exponent and mantissa according to the IEEE Standard for Binary Floating-Point Arithmetic (IEEE 754) [2].

Since the regular integration of the floating-point coprocessor in the processors for personal computers (in the x86 world, this became reality when the 486DX replaced the 486SX in 1992), floating-point arithmetic was available with high performance and became a ubiquitous feature for computer programs.

On systems without a dedicated hardware module for floating-point operations, these have to be emulated by a series of integer operations in software. This is typically the case for embedded microcontrollers, which in most cases do not come with a floating-point module since die size and, therefore, the cost of a microcontroller is strongly increased when a floating-point unit has to be added in hardware. Also, many embedded applications do well without hardware floating-point support. The NIOS II soft core processor [3], for instance, requires around 700 Logic Elements when synthesized as a 32 bit integer processor onto an FPGA (Field-Programmable Gate Array). When adding a floating-point unit, the final design requires around thrice the size of the plain integer design.

Without compiler support, a programmer either has to code the operations manually, use one of the many libraries available for fixed-point operations or use tools like Matlab [4] that are able to generate C code that simulates fixed-point arithmetic.

Existing solutions for fixed-point libraries suffer from one of the following deficiencies: (i) the data types are not standardized, thus it is not possible to reuse code with a different library, (ii) not all required functions are supported, e.g., missing support for trigonometric functions, (iii) if the library is rather complete, the overhead on linking the library to the final program creates a large memory footprint, and (iv) the library is not written in C but in C++ or an architecture-specific assembly language.

Unfortunately, until recently, there was not much compiler support for fixed-point data types and no standard for implementing fixed-point data types. In 2008, ISO issued a standard that is describing the syntax and the data types for fixed-point arithmetic as an extension to the programming language C [5]. However, for particular embedded target systems there is still a lack of compilers that support this standard. Therefore, we have implemented the data types and functions of this standard as a software library that can be used with any standard C compiler.

It is the purpose of this chapter to describe a high-level language implementation of the main parts of the ISO/IEC Standard TR 18037 and evaluate the results by comparison to the standard software floating-point library of the avr-gcc compiler, a compiler for the embedded AVR 8-bit microcontroller series. Our intention is to provide a very generic implementation that can act as a transitional solution for systems where compiler support for the new standard is not yet available as well as a solution for applications with moderate performance requirements. Moreover, the timing behavior of the functions in our library has been thoroughly analyzed on the AVR architecture so that these data supports static Worst Case Execution Time analysis methods [6] for real-time systems on that hardware.

The rest of the chapter is structured as follows:

Section 8.2 reviews some basic properties of fixed-point and floating-point arithmetic. Section 8.3 gives a short introduction to the ISO/IEC 18037 standard. Section 8.4 describes our implementation of a library. Section 8.5 depicts the evaluation results for our library on the AVR architecture. Section 8.6 compares the results to floating-point operation. Section 8.7 concludes the chapter.

8.2 A Closer Look on Fixed-Point Arithmetics

Most programming languages offer only floating-point arithmetic in order to express fractional numbers. Being noticeable exceptions, ADA and COBOL are one of a few programming languages that also natively support fixed-point data types. The reason for this is that floating-point arithmetic comes with the following advantages over fixed-point numbers:

- They approximate real numbers over a relatively wide data range. For example, the float data type in C allows to express numbers between $1.175 \cdot 10^{-38}$ and $3.403 \cdot 10^{38}$. The double data type even supports numbers between $2.225 \cdot 10^{-308}$ and $1.798 \cdot 10^{308}$.
- They provide, except for special situations, like for example very small numbers near zero, a constant relative precision for approximating a real number. The float data type has a precision of $2^{-24} = 5.960 \cdot 10^{-8}$, the double data type has a precision of $2^{-53} = 1.110 \cdot 10^{-16}$.
- The number format is standardized by IEEE Standard for Binary Floating-Point Arithmetic (IEEE 754) to four different precisions: single, double, single-extended, and double-extended. Typically, programming languages and most floating-point hardware support single and double precision [7]. There exists another standard, IEEE 854 [8] that also supports specifying floating-point numbers on the basis 10, which is aligned to representation of numbers by humans. However, unlike IEEE 754, IEEE 854 does not specify how a binary format of a number should look like so that is less applicable in computer programming.

Therefore, most applications will do well in using floating-point numbers. However, there are some points in favor of fixed-point numbers:

- Operations on fixed-point numbers are less complicated than floating-point numbers. This argument is especially of weight, if the target hardware has no special support for floating-point calculations, which is often the case for embedded hardware. Therefore, using fixed-point arithmetic can yield a performance benefit.
- When the numbers to be operated with are in an a priori known order of magnitude, a fixed-point data type that has the radix in an appropriate position can store a number more efficiently than the floating-point data type, since the latter has also to store the flexible exponent. Again, this is an issue for embedded systems with a limited amount of RAM memory.
- Due to the reduced complexity of the fixed-point operations, the resulting code size is by a few kbytes smaller when using fixed-point instead of floating-point numbers. Accordingly to the previous argument, this is an issue for embedded systems with a limited amount of program memory.

If an application requires floating-point arithmetic or does better with fixed-point depends mainly on the data set the application has to deal with. Frantz and Simar [9] discuss this on the example of video and audio processing: While discrete cosine transformations and quantization operations as they appear in video signal processing can be effectively handled using integer operations while audio processing typically uses cascaded filters where each filtering state propagates the error of previous stages. Furthermore, audio signals must retain accuracy even if the signal approaches zero due to the sensitivity of the human ear, which makes audio applications less suitable for fixed-point arithmetic.

Therefore, the employment of fixed-point arithmetic can be advantageous for some embedded applications where fractional numbers are required, but floating-point operations would be too expensive in terms of hardware cost or processing time.

8.3 Floating-Point Extensions According to ISO/IEC TR 18037:2008

Because fixed-point operations are commonly used for microcontrollers, the ISO/IEC has summarized some guidelines and suggestions in a technical report named "Extensions for the programming language C to support embedded processors" [5], which defines some guidelines for including fixed-point data type support into C compilers. This includes data types,

#pragma directives, constants, function names, and some mathematical conventions.

Although the standard is intended to describe fixed-point extensions for C compilers, we have decided to use those guidelines for the implementation of an external C library.

Having the real time and code size limitations in mind, we decided to implement a reasonable subset of the data types defined by the standard. The overall set of data types and the implemented types are shown in Table 8.1. The numbers indicate the bits that are available for representing the integer part and the fractional part. Note that the definition of our implementation exceeds the number of specified bits, which is still in conformance with the standard, since the given values specify the minimum number of bits for the data types.

Table 8.1 Fractional data types according to ISO/IEC TR 18037

ISO/IEC Definition		Implemented	
signed short _Fract	s.7		
signed _Fract	s.15		
signed long _Fract	s.23		
signed short _Accum	s4.7	_sAccum	s7.8
signed _Accum	s4.15	_Accum	s15.16
signed long _Accum	s4.23	_lAccum	s7.24

FX_ACCUM_OVERFLOW defines the overflow behavior of the Accum data types. When set to SAT, saturation is enabled; this means that when an overflow occurs, the result is either the minimal or the maximal possible value of the data type. This behavior often means a significant loss of speed and further increases code size, so the #pragma is normally set to DEFAULT, which is the other possible value. Since we cannot implement #pragmas in a library, this behavior can be set with a #define FX_ACCUM_OVERFLOW before the library header file is included.

FX_FRACT_OVERFLOW is the same for the Fract data types. Since we have none of them implemented, this #pragma is not used in our library.

FX_FULL_PRECISION forces the implementation to gain maximum precision, by allowing a maximum error of one ULP (Unit in the Last Place) of the result. However, for particular problems, a precision of 2 ULPs on multiplication and division operations is sufficient, which enables optimizations towards execution speed and code size. For our implementation the FX_FULL_PRECISION switch is not implemented, instead the precision has been predicted separately for each function.

Almost any meaningful data type handling and some low level arithmetic functions are defined through naming conventions and behavior descriptions. There is one version for each data type respectively several for conversion and mixed type functions. Except for their parameters they differ also by some trailing and/or leading characters which describe the type of the parameters respectively the result.

8.4 Library Implementation

The primary goal of this work was to provide a fixed-point library especially for use with Atmel 8 bit processors in combination with real time applications. So, performance and performance predictability were strong requirements for the design. Also flash memory was very limited, therefore small code size was desired.

To optimize code for size and speed, every function was implemented as accurate as possible, trying to keep it mathematically fast and simple (thus reducing code size). Apart from optimizing the overall code size of the library, each function has been compiled into a separated object file that is only linked to the final program if the function was used.

A main decision that was made refers to the data types. The ISO/IEC paper recommends both the _Fract and the _Accum type. The difference between those two types is only the lack of integral bits in the _Fract type while not increasing the number of fractional bits, so we decided to implement the _Accum type. To further limit the complexity of the implementation, we only implemented two subtypes of the _Accum type. Although the two data types should be named _Accum and long _Accum, there is a problem with the name of the second type. As we use typedef to define the type, the name of the data type must not have blanks in it. So we decided to call it _lAccum, which should be kept in mind when comparing AVRfix with the ISO/IEC specification.

Both types are signed and held in a 32 bit container (signed long). While _Accum has 15 integral and 16 fractional bits, _lAccum has only 7 integral bits but therefore 24 fractional bits. Because we use the long data type as container, addition and subtraction are working implicitly by using integer arithmetic as long as _Accum and _lAccum are not mixed. Therefore, addition and subtraction require no additional function in the library.

Overloading of operators is not supported in ANSI-C, so for example a multiplication needs to be done by a function call or a macro. While function calls produce some overhead on runtime, the use of macros increases code size. Most operations except conversion functions are implemented as functions. Comparison functions are working as long as the data types are the same; casting has no effect for _Accum and _lAccum. If a comparison between a long and an _Accum is needed, one (or both) of the variables needs to be converted before the comparison can be done. The same approach is needed for assignments.

To meet requirements of code size and execution speed, the FX_FULL_PRECISION switch has not been implemented. Instead, the expected precision has been evaluated and documented separately for each function in the project documentation [10]. This evaluation includes also sophisticated math functions such as trigonometric functions and square root. The library is completely written in C and has been tested and evaluated with the established compilers avr-gcc 3.3.2 and the Microsoft Visual Studio IDE 6.0.

In reference to the ISO/IEC report, the naming conventions are used accordingly whenever possible, meaning that for _Accum a *k*, for _lAccum *lk* and for _sAccum *sk* is used as suffix to the function name to indicate the type of the parameters. The type of the return value is indicated by a letter before the function name. No letter suggests _Accum, an *l* means that the return value is of type _lAccum, and an *s* refers to _sAccum.

For example, the multiplication function that multiplies two _Accum values and returns an _Accum value, is named mulk. The multiplication function that multiplies two _lAccum values and returns an _lAccum value, is named lmullk.

The ISO/IEC paper specifies the FX_ACCUM_OVERFLOW flag, which defines the behavior if an overflow occurs. If it is set to saturation (SAT), the value will be either the maximum or minimum possible value if an overflow occurs. By default, an overflow will give an undefined result. While in the ISO/IEC paper this flag is defined as a #pragma directive, we needed to use a #define for the FX_ACCUM_OVERFLOW flag. Independent from this flag the behavior can be achieved by calling the respective version of the function directly. If a function provides both behaviors, there exist two functions which have either S for saturation or D for default behavior as trailing character after the function name. So one can attach an S to a function name to force saturation behavior or a D to force the default behavior (resulting in e.g. mulkD or mulkS for the two versions of mulk) if the function provides two different behaviors.

Apart from the four arithmetic basic operations, the library support also operations such as square root, logarithmic and trigonometric functions. For the latter we have decided to use the CORDIC approximation [11] instead of a Taylor series, because it saves considerable program memory when requiring sine and cosine (or, consequently, tangens) in the same program while achieving almost the performance of the Taylor version.

8.5 Evaluation

For tests and benchmarks we used an evaluation board equipped with an Atmel ATMEGA 16, providing 16 MHz clock, 16 Kb flash memory and 2 Kb SRAM. For evaluating the correctness of the calculations done by the library, we tried to cover all meaningful calculations. To speed up this brute force approach, we mainly did this on a PC and compared the result with either results from 64 bit integer calculations or precalculated reference results. The reference results have been created with the statistical computing environment R for a meaningful range. For example, the meaningful range for sine and cosine is from zero to two times Pi, meaning for an _Accum parameter, that 411774 calculations and comparisons had to be done. For functions that have no fixed execution time, the execution time over parameter is recorded and visualized via gnuplot.

8.5.1 Evaluation on the Microcontroller

To measure execution time and verify the calculation results, we wrote a small microcontroller program. To measure execution speed, we use the 16 bit timer. The counter is reset to zero, the function is called and the counter value is fetched afterwards. The execution time, parameters and result is then transmitted via UART. To speed up transmission, a high bit rate is used and the data is sent binary, so a conversion was needed to plot the data in gnuplot.

8.5.2 Accuracy Test

To test the accuracy of the implemented functions, we compared the output values with precise 64-bit calculations for the _Accum and _lAccum data type (respectively with precise 64-bit calculations for the _sAccum data type) for addition/subtraction, multiplication and division. For higher mathematical operation pre-calculated reference values have been used. The accuracy test was done on a PC as we use regular C code and the execution is much faster as on the microcontroller. We assumed equality of the output after some calculations done on both, the PC and the microcontroller. The established Microsoft Visual C++ 98 environment was used as the reference compiler.

8.5.2.1 Multiplication and Division

To test multiplication and division, we simply treated the _Accum and _lAccum values as signed 64-bit integer values, repeated the calculations with 64-bit accuracy and compared the results. Testing has been done completely for the _sAccum type. For the other data types, extensive testing including the critical value pairs has been performed.

For a multiplication $x \cdot y$, all values $|x| > (2^{\wedge i+f} - 1) \cdot 2^{-f} / y$ will lead to an overflow for $y > 1$, with i being the number of integral bits an f being the number of fractional bits of the data type. For a division x/y, all values $x > (2^{i+f} - 1) \cdot 2^{-f} \cdot y$ will lead to an overflow. According to the data type, this is only a limitation for x if $y < 1$.

Extensive testing of the functions showed the following results:

- The _sAccum functions have a maximum error of 0 for multiplication and division, both tested with default and saturation behavior.
- The _Accum functions have a maximum error of 2^{-16} for multiplication and well-defined division calculations, both tested with default and saturation behavior.
- The _lAccum functions have a maximum error of 2^{-24} for multiplication and well-defined division calculations with default behavior. For saturation behavior, the maximum error was 2^{-23}.

8.5.2.2 Extended and Trigonometric Functions

For extended and trigonometric functions (e.g. sine/cosine, logarithm etc.), the comparison values were provided by R from the R Foundation, a statistical calculation environment. Most sophisticated functions have an error behavior that strongly depends on the input. Thus regarding the error over the whole input range makes sense. For example, Figure 8.1 depicts the error function for the _Accum square root function. An exhaustive evaluation of all functions can be found in the project documentation[10].

8.5.3 Performance Testing

Our first attempt to test performance of our implementation was to use the destination device, an Atmel ATMEGA 16, but as its maximum speed is 16 MHz and the serial port is a very slow transmission system, we decided to go a different way. We implemented a very simple simulator to test the performance on a PC.

8.5.3.1 The Disassembler & Simulator Creator (DsimC

The *Disassembler & Simulator Creator (DsimC)* is a little Java Application that disassembles an .srec-file for an Atmel ATMEGA16 and transforms each instruction into a piece of C code. This code can be compiled and executed on a PC instead of downloading and executing the original code on the microcontroller.

This was possible, because the ATMEGA16 has no caches or other elements that make code execution times indeterministic, but only a two-stage pipeline with very low effect on execution time. So each hardware instruction is expanded to a group of C code instructions which performs an equivalent operation, maintains the virtual status register flags and increments a tick counter which furthermore can be used to determine the performance of the library functions. In addition, every write to the UART Data Register (UDR) results in a file output operation, which gives us a very high speed up. As assumed, the maintenance of the status register flags turned out to be most expensive, resulting in a simulation speed of only about 25–30 times faster than on the ATMEGA16 when using a Pentium-M with 2 GHz. This seems to be a good speedup, but most of it comes from the serial port implementation.

When we compared the performance values calculated by our simulations with values we determined on the microcontroller, we noticed a slight drift. It turned out that the simulator counts too many ticks under certain conditions, resulting in a few ticks more per function call, if ever. But when we tried to isolate the operations causing this drift, it turned out to be very tricky because of lack of an in-circuit debugger for the microcontroller we would have to flash the target many times to reduce the code range in which the drift appears. In our analysis we have noticed that the drift is only in one direction, if ever. Fortunately, the simulator never gives fewer ticks than it would take on the microcontroller, so this is sufficient to get guaranteed worst case execution time values. However, by taking the discrepancies between simulator and real hardware into account, tighter WCET values would be possible.

8.6 Comparison

The traditional way for fractional computing is the usage of floating-point operations, for which a various number of libraries exist. We compared our fixed-point library with the floating-point library (libm) that comes with the avr-gcc bundle.

8.6.1 Accuracy

All data types compared here (float, double, _Accum and _lAccum) reside in a 32-bit container. But the floating-point data types have to separate the container for exponent and mantissa, while the fixed-point data types have the whole container for the sign bit, the integral bits and the fractional bits. So, within the fixed-point range $(2^{31}-1) \cdot 2^{-16}$ and $(2^{31}-1) \cdot 2^{-24}$, respectively, the _Accum and _lAccum types are more accurate as long as the value to be expressed matches the fixed-point range in its order of magnitude. The _sAccum data type has clearly a lower accuracy than the floating-point data types since it resides in a 16-bit container only.

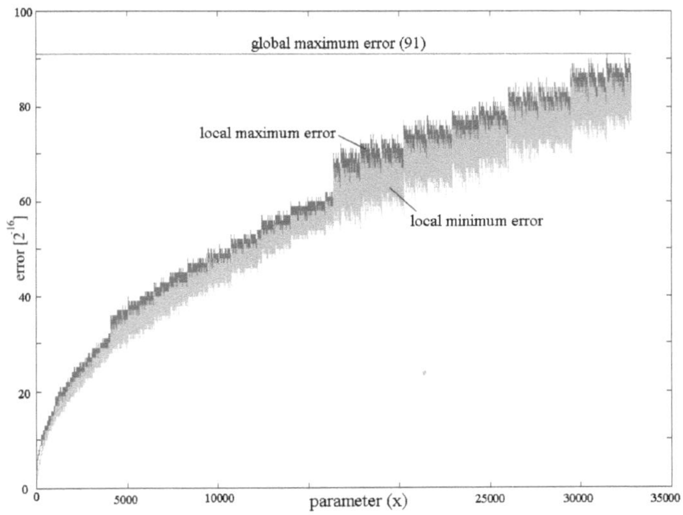

Fig. 8.1 Accuracy distribution for *sqrk*

8.6.2 Addition and Subtraction

The fixed-point addition and subtraction operations use the same instructions as normal integer operations, so they really make the cut over floating-point addition and subtraction (as shown in Table 8.2).

8.6.3 Multiplication

Compared to the multiplication functions mulkD, mulkS, lmullkD and lmullkS the average performance of the double operation is higher, while the WCET of the fixed-point multiplication functions is better. This is due to the fact that the double data type is only implemented in 32 bit by the avr-gcc, thus a double multiplication involves a 23 bit multiplication of the mantissa and an addition of the exponent. In addition, the floating-point library has been optimized at assembly code level for average performance, which explains the results (cf. Table 8.3).

Table 8.2 Performance comparison for addition operations

Data type	Execution time in ticks
Double	74–80
_sAccum	14
_Accum and _lAccum	23

Table 8.3 Performance comparison of multiplication operations

	Execution time in ticks	
Data type	Default	Saturated
double	53–2851	-
_sAccum	79–82	92–95
_Accum	337–350	215–359
_lAccum	594–596	198–742

8.6.4 Division

Compared to the division functions divkD, divkS, ldivlkD and ldivlkS the minimum, average and maximum performance of the double operation is in general better, which can be seen in the overview given in Table 8.4. As a consequence, if possible, fixed-point divisions should be avoided for performance reasons.

Table 8.4 Performance comparison of division operations

	Execution time in ticks	
Data type	Default	Saturated
double	66–1385	-
_sAccum	634–711	650–727
_Accum	820–1291	853–1386
_lAccum	876–1405	862–1416

8.6.5 Floating-Point Code Size

We have measured the code size for using the floating-point functions provided by the compiler. Using a simple addition, for example adds about 1740 bytes in code size. To cover the basic arithmetic operations about 3k of Flash ROM are needed. In contrast, when using AVRFix and the datatype _Accum with default behavior, only 758 bytes are needed. For _lAccum 848 bytes and for _sAccum only 260 bytes are needed. Thus, AVRFix has a clear advantage in code size compared to floating-point operations.

8.7 Conclusion

The contributions in this chapter are the implementation and evaluation of a generic fixed-point library based on the ISO/IEC Standard TR 18037. The documentation [10] and the source code is available as open source.

The fixed-point library contains not only basic mathematical functions and conversions but also more sophisticated operations such as square root, logarithmic and trigonometric functions. The linking model allows having only the used functions in the final assembler code, which saves considerable program memory over monolithic libraries.

We have performed exact performance measurements for a specific target architecture, the Atmel AVR with avr-gcc compiler. The results from this analysis can be used for static WCET analysis and optimization of execution time and code size, which is of special interest for embedded application on low-cost microcontrollers with few resources.

Addition and subtraction are generally by a factor of 3–5 faster than floating-point operations. The fixed-point multiplication and division have worse average performance, but a better WCET than the floating-point operations for most data types. Moreover, since the library has been written in C, there is also room for hardware-specific optimizations of the library, e.g., by using inline assembler functions for time-critical parts. Regarding code size the fixed-point operations are clearly in favor.

If only addition, subtraction and multiplication is needed or a small data type like _sAccum is sufficient, the use of fixed-point operations can clearly be favored from the viewpoints of speed, WCET, and code size. Sophisticated functions like trigonometric and exponential functions are slower than the fixed-point versions, but require less program memory, which makes the fixed-point implementation attractive for projects on small embedded microcontrollers where program memory becomes the main limiting factor. The optional saturation behavior is a nice feature which cannot easily be reproduced by floating-point calculations and small code size may be a decisive advantage.

In the future, we expect the fixed point standard to be supported by embedded compilers. The current version of gcc v4.3.1 supports the fixed point extensions only for the MIPS target. Being integrated into the compiler, we expect an increase in performance for compiler-supported fixed point arithmetic in comparison to our library. Until there is sufficient compiler support, our library can be a transitional solution that allows developers to use fixed-point arithmetic.

Acknowledgement This work was supported in part by the Austrian FWF project TTCAR under contract No. P18060-N04.

References

1. ISO/IEC, Programming Languages – C, approved by ANSI Accredited Standards Committee, ISO/IEC 9899:1999, December, 1999.
2. IEEE, Standard for Binary Floating-Point Arithmetic, ANSI/IEEE Std 754-1985; IEC-60559:1989, 1985.
3. Altera Corporation, USA. Nios II Processor Reference Handbook Version 7.0, 2007
4. D.P. Magee, Matlab extensions for the development, testing and verification of real-time DSP software, In Proceedings of the 42nd Annual Conference on Design Automation, pages 603–606, San Diego, CA, USA, 2005.
5. ISO/IEC, Programming languages – C – Extensions to support embedded processors, ISO/IEC TR 18037:2008, JTC 1/SC 22, 2008.
6. P. Puschner, Worst-case execution time analysis at low cost, Control Engineering Practice, 6:129–135, January 1998.
7. D. Goldberg, What every computer scientist should know about floating-point arithmetic, ACM Computing Surveys, 23(1):5–48, March, 1991.
8. IEEE, Standard for Radix-independent Floating-point Arithmetic, ANSI/IEEE Std 854-1987, October 1987.
9. G. Frantz and R. Simar, Comparing fixed- and floating-point DSPs. Texas Instruments, Dallas, TX, USA, 2004. White paper available at http://ocus.ti.com/lit/ml /spry061/spry061.pdf.
10. M. Rosenblattl and A. Wolf, Fixed-point library according to ISO/IEC standard DTR 18037 for Atmel AVR processors, Bachelor's thesis, Vienna University of Technology, Vienna, Austria, 2007. http://sourceforge.net/projects/avrfix.
11. J.E. Volder, The CORDIC trigonometric computing technique. IRE Transactions on Electronic Computers, EC-8(3), 9, 1959.

Part III

Middleware Platforms

Chapter 9

A Home E-Health System for Dependent People Based on OSGI

Jaime Martín[1], Ralf Seepold[1], Natividad Martínez Madrid[1], Juan Antonio Álvarez[2], Alejandro Fernández-Montes[2] and Juan Antonio Ortega[2]
[1]*Universidad Carlos III de Madrid, Spain;*
[2]*Universidad de Sevilla, Spain*
natividad.martínez@uc3m.es, ralf.seepold@uc3m.es, jortega@us.es, alejandro.fdez@gmail.com, jaalvarez@us.es

Abstract This chapter presents a e-health system for dependent people installed in a home environment. After reviewing the state of art in e-health applications and technologies several limitations have been detected because many solutions are proprietary and lack interoperability. The developed home e-health system provides an architecture capable to integrate different telecare services in a smart home gateway hardware independent from the application layer. We propose a rule system to define users' behavior and monitor relevant events. Two example systems have been implemented to monitor patients. A data model for the e-health platform is described as well.

Keywords e-health, telemedicine, telecare, dependent people, orientation, home automation, OSGi, HL7.

9.1 Introduction

The World Health Organization defines the term e-health as the relations between institutions, public health, e-learning, remote monitoring, telephone assistance, domiciliary care and any other system of remote medicine care. Telecare is the part of e-health that offers remote care of dependent (elderly or disabled people), providing the care and reassurance needed to allow them to live in their own homes.

Integration of healthcare key actors is required to offer quality in service but e-health systems often lack adequate interoperability or integration of social aspects; the results is a slow down in acceptance and usage of these

N. Martínez Madrid, R.E.D. Seepold (eds.), *Intelligent Technical Systems*, Lecture Notes in Electrical Engineering 38, DOI 978-1-4020-9823-9_9,

systems. Integrating Information and Communication Technologies (ICT) (like for example telehomecare) in care, living and wellness is a citizens demand that it should be provide at affordable cost [1].

An example of e-health services integration challenge is to achieve communicate to dependent people with relatives and medical people, integrating movement monitorization and orientation together with electronic medical record transmission. Consequently, multimedia communication and tracking people technologies should be seamlessly integrated in the e-health service. Residential Gateways (RGW) enable telecom companies to provide applications in a home environment via a platform to manage several services remotely. The OSGi (formerly known as the Open Services Gateway initiative, www.osgi.org) specification provides an architecture for remote control of a platform. OSGi also includes support for the whole life-cycle of services (e.g. start, stop, etc.). This chapter proposes a platform for a home e-health system based in OSGi that provides capabilities to integrate different telecare, telemedicine and orientation support services for dependent people in the same RGW.

In Section 9.2, the state of the art of telemedicine, telecare and electronic health record standards is reviewed. Section 9.3 presents the new proposal by describing two patient monitorization examples, a production system, use cases and the data model. Finally, Section 9.4 sums up the results so far and describes some future work.

9.2 State of Art

9.2.1 Related Research on Telecare and Telemedicine

Several residential telecare and telemedicine platforms approaches are presented in literature. Bobbie [2] describes an electronic-prescription system for home-based telemedicine using the OSGi framework. This article describes a health-prescription application running on a smart card that communicates with a Personal Digital Assistant (PDA). It uses OSGi as a central coordinating point between different devices. The OSGi environment is aimed to allow intercommunication between the card reader, the patient's PDA application and other devices. However, it lacks a detailed description about how the system is implemented and how security needs are reflected in the application implementation that are specific in a medical environment.

The Service-Oriented Agent Architecture described in another approach [3] enables healthcare services providers to support telecardiology services on demand. It proposes a runtime unit of telemedicine agents to permit services to be managed remotely. The system consists of an agent unit, which includes a vital signals acquisition module; it can acquire ECG (electrocardiogram) data and forward ECG data to a medical service center. It uses Web Services to communicate to all services via XML in multiple platforms but it does not mention any EHR standard.

Some Spanish research projects have presented a generic architecture to e-health system based on middleware components [4]. The goal is to provide a service platform for any kind of e-health application. These generic architectures should allow a quick development of new services using the existing infrastructure. Currently, the HL7 standard is ruled out because it lacks interoperability with other systems and EN 13606 is proposed to represent and to transmit clinic information. Currently, a proposal of an IEEE 11073 platform for healthcare tele-monitoring has been published [5]. This work identifies a set of use cases relevant for a personal monitoring scenario and it identifies related features and functionalities.

9.2.2 E-Health Technologies

Health patient data must be transmitted and saved in a standard format supported by all involved systems. An electronic health record (EHR) refers to an individual patient's medical record in digital format. An EHR standards comparative study [6] describes HL7 and EN 13606 standards. To transmit medical information between devices IEEE 11073 has been developed. A brief overview of these technologies is presented in the following three sub-sections.

9.2.2.1 HL7

Health Level Seven (HL7) [7,8] is a widely applied protocol to exchange clinical data. Several versions are been developed by the HL7 organization, part of American National Standard Institution and founded in 1987. Version 3 is beyond the scope of this chapter because it is a complex standard and there is no a stable version available [9].

The HL7 refers to seventh OSI layer (application) although it also specifies a layer 6 presentation protocol made up of its own abstract message format and encoding rules. Concerning the lower layers, like session and transport services, it is rather vague because the HL7 authors' intention was to support a wide variety of systems. The underlying HL7 operational model is a client-server system. For example, when a patient is admitted to a hospital, the admission system will propagate HL7 admission messages to an appropriate subsystem. An HL7 message always contains all the information required to complete a transaction and it is encoded in HL7 rules. Essentially, all information is transmitted in ASCII plain text. The standard allows defining site-specific extensions segments, like message extensions to exchange data with an appointment system. However, the use of these extensions can prompt serious interoperability problems. Moreover, HL7 lacks a specific methodology to generate messages and it is not clear how the structural relations between fields are defined.

9.2.2.2 EN 13606

Health informatics – Electronic Health Record Communication standard (EN 13606) is a European official standard of CEN (European Committee for Standardization) and an approved ISO standard. The overall goal is define a rigorous and stable information architecture for communicating part or all of the EHR of a patient. It is based on the HL7 RIM (Reference Information Model) from HL7 v3, a set of data type definitions harmonized between HL7 and CEN, the EHR Domain Information Model (DMIM) and a bunch of RMIMs dedicated to certain structures and functionalities.

EN 13606 is flexible to represent the information structures transmitted thanks to the archetypes, a knowledge representation of the clinic information domain. Moreover, it is robust with respect to changes in the specifications because changes in the archetypes do not provoke a change in the underlying system.

The openEHR framework (www.openehr.org) is compliant to the EN 13606 and it is used in commercial systems throughout the world.

9.2.2.3 ISO/IEEE 11073

A brief description of novel standards ISO/IEEE 11073, often also referred to Medical Information Bus (MIB), or x73 standards, for personal telemedicine systems interoperability can be found in [10]. The goal is to enable medical devices to interconnect and interoperate with other medical devices. These standards cover the upper OSI layers and use well-known IEEE standards like Bluetooth (802.15.1) or WLAN (802.11) in lower layers. Part of x73 standards focus on point-of-care medical devices communication which are mainly designed for acute monitoring and treatment application in a particular diagnostic, bed or treatment area in the hospital domain like Intensive Care Unit (ICU). Several of x73 standards series are currently drafts and new projects are under development; first prototype implementations are available in industry.

The standard is based on an object-oriented system management paradigm. A numeric code set identifies every item that is communicating. This is more efficient than HL7 because it uses binary instead of plain text data. The key objectives for clinical domain applications addressed by the standard are real-time plug & play interoperability and frequent network reconfiguration. Special attention has been paid to reduce implementation complexity and computational burden of devices. For wireless devices, transmission power and transmission time could be reduced.

9.3 E-Health Service Proposal for Dependent People

9.3.1 Overview

Our proposal attempts to integrate several smart home services to provide a scalable and interoperability e-health solution. The system is divided into three basic subsystems: domotic, multimedia and e-health subsystem. In the domotic subsystem different devices can co-exist in each subsystem connected by wire or wireless to a residential gateway (RGW) based on an embedded OSGi framework. Blood-pressure monitor and a pair of scales are examples of integrated devices in the medical network.

A domotic environment based on a Lonwork[a] network typically includes sensors and actuators, for example light sensors or blind motor. A Lonworks platform provides a reliable and open protocol accepted as a standard for control networking. As we see in data model, a RGW can obtain some useful environment variables from automation home network so when a relevant event occurs, an alert or alarm is reported in the RGW. This event can be sent to relatives or medical people. The multimedia network typically includes a television, an IP camera or a webcam with microphone, necessary for dependent person to communicate. The RGW is able to physically interconnect all required networks and devices, and to host different services which can be managed remotely by the e-health or access provider. Multimedia services, like SIP audio/videoconference are provided to communicate during the medical tele-visit (cf. Fig. 9.1).

[a] http://en.wikipedia.org/wiki/LonWorks

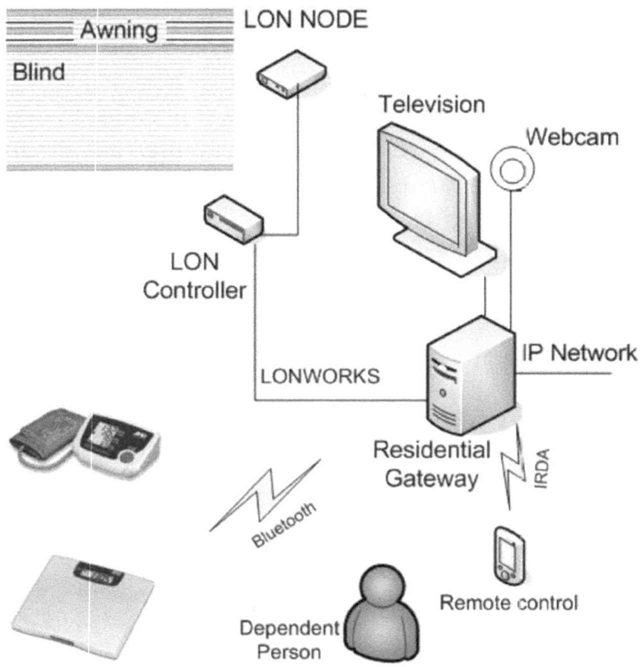

Fig. 9.1 Smart Home overview, with automation, multimedia and medical devices

9.3.2 Movement Monitorization and Orientation

In this section we propose two systems to monitor patients. The first one monitors outdoors trips and the second one indoor activity.

The first uses an inference engine that does not require users to explicitly provide information about the start or end points of their journeys; instead this information is learned from users' past behavior. The initial requirement is the tracking of frequent trips of the monitored users. In order to test our system, a user carried a GPS logger with him continuously for 24 hours a day for 30 days. GPS data was filtered and segmented to obtain the user's trips. Then the GPS trips with spatial and temporal data were clustered in routes (only spatial information) using a dendrogram clustering tested by Froehlich and Krumm at Microsoft [11]. The routes were used to make a supervised learning with new trips. Figure 9.2 shows the routes obtained from one user.

Using similarities that measure point to point distances between a new trip and the existing routes and a very early classification was done about the route and the user's goal when the trip is beginning.

The trained model correctly identifies more common places and whole routes, allowing the detection of lost or disorientation situations in open areas.

Fig. 9.2 A example of GPS tracking

When a set of GPS locations are received and all of them are located out of the security area of the routes, different policies could be selected to prevent dangerous situations:

- If the user is a dependent person or his abilities handling a smartphone are low, the first option is to send an automatically generated SMS send to previously selected people showing the exact location of the user. If an Internet connection is available, the text of the SMS is an address obtained from a Web Service that provides the close address to a latitude-longitude pair (reverse geocoding). If not, the text is only the latitude-longitude pair.

- Another policy is a reorientation service that helps the user to reach the closest point from a frequent route and then guide him to a frequent place (he could choose one from photographs of the common ones). In a previous work [13] we attempted to design a reorientation application to Alzheimer patients but the lack of families that allow a relative that suffers that disease to go out alone and the difficulties to manage a device like the HTC P3300 by this kind of people in a crisis situation make us change the interaction and the potential users. In [14] a similar application was explained but the users' audience was different (mild cognitive disabilities).

Due to the fact that users usually are indoors, the system proposed was not complete to monitor and help them. To improve the usefulness, a second system has been designed using the smartphone and a device called Alive Heart Monitor [15] (cf. Fig. 9.3) composed by an ECG and a tri-axial accelerometer. The weight (55gr with battery), measurements (90×40×16 mm), and operating live without change the battery and sending the data wireless (48 hours) make it perfect to be worn by the user. The only problem is that two liquid gel electrodes must be placed on the user's chest to obtain high quality ECG signal. This is normally bothering the user for a long term monitoring.

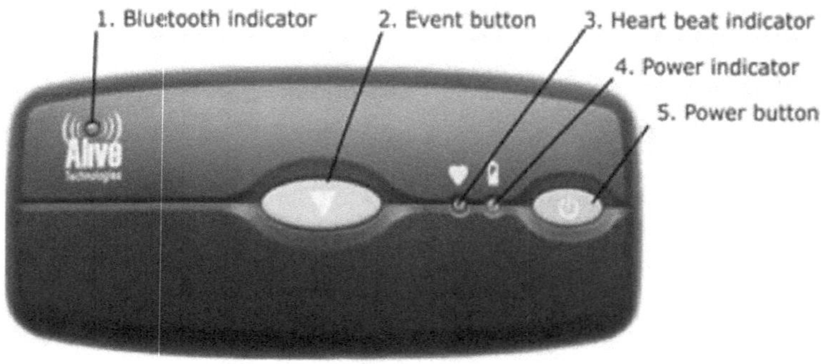

Fig. 9.3 Health monitor

The ECG data helps to detects potential heart illness and the heart rate. The accelerometer data depending on the position of the device helps to classify the activities of the user (running, walking, standing up, sitting down) and other risky patterns like falls (a big acceleration followed by a sudden stop and a body change of orientation).

A rule-alert system was designed to send automatically a SMS to the relatives; an OSGi-Server stores vital constants to be queried by authorized users.

9.3.3 Production System

A rule-based system to control and monitorize the user's behavior is presented here. This is a production system formed by a facts base and rules base as showed in Fig. 9.4. This has been implemented as a rule engine and scripting environment[b] written in Java language by Jess.

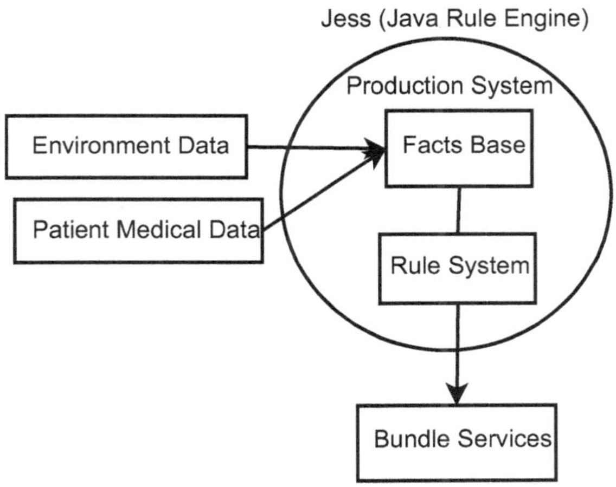

Fig. 9.4 Production System for e-Health system

In Table 9.1 we show an example of Rules Base for our e-Health platform with health and automatic rules based on a health monitor and other sensors.

Table 9.1 Example of rules base for e-health system

Condition	Action
High Blood Pressure Rate	Call to health professionals
High Blood Pressure Rate	Call to relatives
Low Light Level	Rise up blind

In certain cases, we need a priority range for these rules because it is possible that several rules apply at the same time. Priority and urgency

[b] http://herzberg.ca.sandia.gov

issues for a mobile care system are treated in [16]. Alert mechanisms in actions should have a priority level because some situations could be critical or very critical and they need a faster response. For example, if the patient has a lot of pain he can contact the assistant before relatives are informed. The assistant can then decided if an ambulance is required to attend the patient.

9.3.4 Use Cases

We can identify some important use case of the platform. A general administrator controls the RGW but can appear that there are different administrators for each subsystem like an e-health admin, which is allowed to configure e-health devices only. A solution to separate different RGW administration by virtualization is described in [17]. Every service offered by a device is part of a scene, i.e. a set of pre-established services by the admin. Users without administration permission, like relatives, friends, assistants or dependent person can directly customize the devices to adapt them according to their preferences. For example, a relative of a dependent person can set the hours that a blind is open to allow illuminate the room during the periodic blood pressure check.

Telecare with assistant people of the patient/dependent person can be another use case, like it is shown in Fig. 9.5.

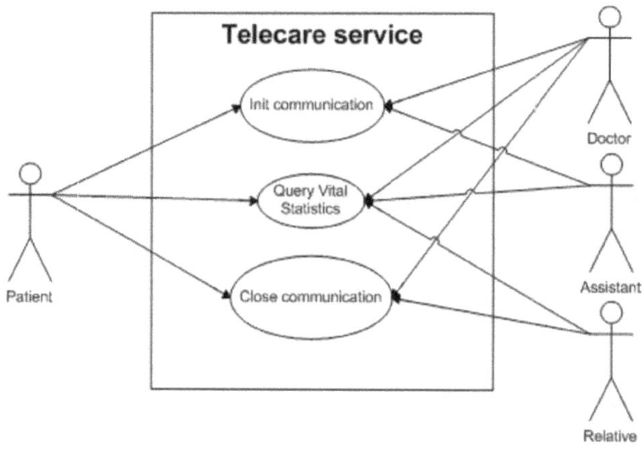

Fig. 9.5 Telecare use case

This is often organized without considering a communication with relatives and friends. The problem is that the patient usually prefers to contact first of all his relatives and friends if they need anything. According to several studies, dependent people are reluctant to use many health care services because they do not personally know the operator or the contact person in the service centre. So an objective of this work is to integrate relatives and friends into the healthcare service provision, and thus increase the usability of the system. For example, in this use case the assistant initiates a SIP video call with the patient and with a relative. He checks remotely the vital statistics like weight or blood pressure thanks to health care devices at the patient home. The wireless devices are connected via x73 standards (or proprietary protocols) to RGW which recovers the medical information, processes data and saves them in HL7 format. Then, the data can be processed by the medical information system and displayed to the doctor in his computer.

9.3.5 Data Model

The data model designed is divided into the user management on one side and the device management on the other side. An overview of data model is shown in Fig. 9.6.

A generic User entity saves basic data, like name, surname, address, etc. Defined attributes in this schema should be compatible with HL7 PID (Patient Identifier Segment) fields. For example, Spanish second surname must be matched in Mother's Maiden Name (XPN) following the HL7 Spain recommendations (www.hl7spain.org). Relative, Assistant, Doctor and Administrator are entities which have different attributes and different roles according to their permissions. A role defines a permission to access to data or devices. In this manner, the user type definition is separated from privilege definition. The administrator user has total control over the full system in case he is the RGW admin or only partially in case he is member of another admin type, like the e-health admin. Automation, multimedia and medical devices have a link address and/or IP address, status, service list and several configuration variables. The RGW can save some parameters of a Lonworks node like for example its ID or its network variables to monitor the status of patient and his environment.

The doctor is allowed to access basic and medical data of the assigned patients. The assistant entity can represent a nurse or a social assistant. Information about medical visits, like diagnostic, reason and date, is saved in Visit entity. If a doctor orders a treatment, important dates and status are saved in Treatment. There is medicine administration sometimes, so frequency, medicine name and comments can be annotated in Medicine. A user (doctor, assistant or patient) can add remainders associated to a treatment, like exercises or medical administration remainders.

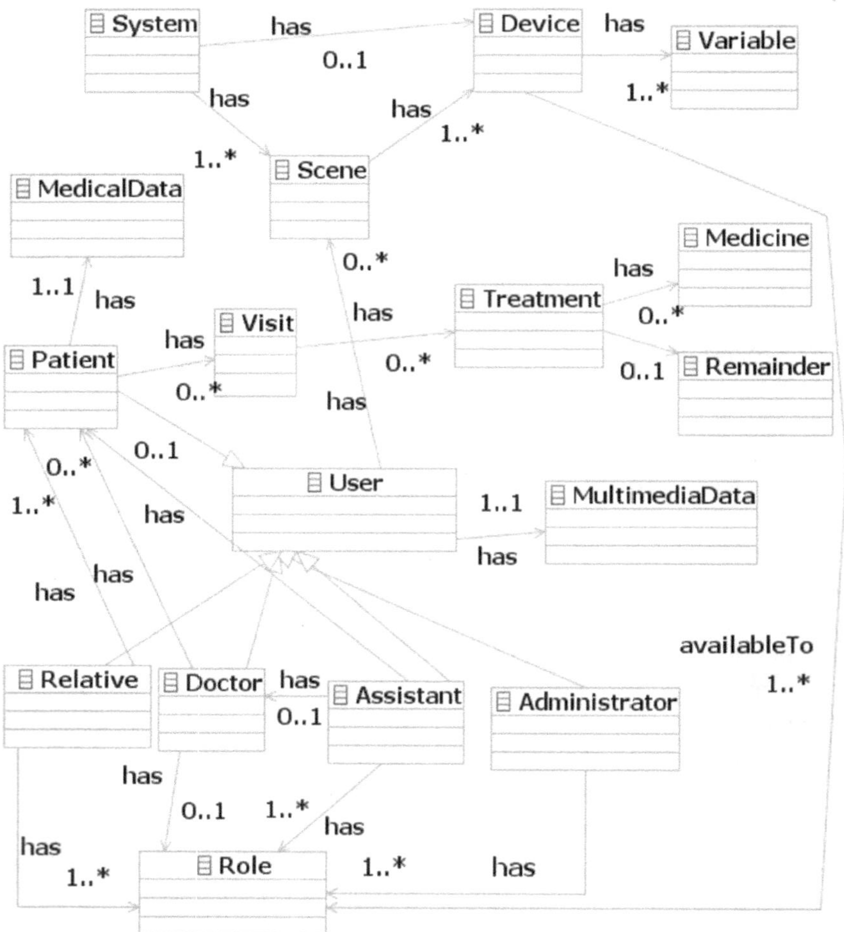

Fig. 9.6 Platform Data Model

9.4 Conclusions and Future Work

A new e-health system architecture has been presented that tries to integrate all relevant members into a common scenario. The dependent person, the medical personal and relatives (or friends) are integrated as members of the platform. Each entity has individual duties following the general objective to enhance the quality of life for the dependent person. Besides this important goal, the technical infrastructure has been developed to seamlessly integrate technical devices. Existing but also extended versions of standards are used and proposed in order to increase efficiency and usability in realistic scenarios. Future work will provide a basic implementation and test of use cases..

Acknowledgements This research is supported by the MEC I+D project InCare-FAMENET. Ref: TSI2006-13390-C02-02.

References

1. EHTEL Force. Sustainable Telemedicine Task. Sustainable Telemedicine: paradigms for future-proof healthcare. A brief Paper. European Health Telematics Association (EHTEL), 2008.
2. P.O. Bobbie, S.H. Ramisetty, A. Yussiff, and S. Pujari. Designing an Embedded Electronic-Prescription Application for Home-Based Telemedicine Using OSGi Framework. In Embedded Systems and Applications, H.R. Arabnia and L.T. Yang, eds., 16–21, CSREA Press, 2003.
3. Y. Chen, and C. Huang. A Service-Oriented Agent Architecture to Support Telecardiology Services on Demand. Journal of Medical and Biological Engineering 25(2), 2005.
4. P. de Toledo, A. Muñoz, J.A. Maldonado, E. Hernando, R. Somolinos, P. Crespo, E. Gómez, F. del Pozo, M. Robles, and J.A. Fragua, Arquitectura genérica para sistemas de e-salud basada en componentes middleware. In Libro de Actas del XXIII Congreso Anual de la Sociedad Española de Ingeniería Biomédica (CASEIB'05), 29–32, 2005.
5. M. Galarraga, I. Martinez, L. Serrano, P. de Toledo, J. Escayolan, J. Fernandez, S. Jimenez-Fernandez, S. Led, M. Martinez-Espronceda, E. Viruete, and J. Garcia. Proposal of an ISO/IEEE11073 Platform for Healthcare Telemonitoring: Plug-and-Play Solution with new Use Cases. In Engineering in Medicine and Biology Society, 2007. EMBS 2007. 29th Annual International Conference of the IEEE, 6709–6712, 2007.
6. B. Blobel, and P. Pharow. EHR Standards-A Comparative Study. In Medical And Care Compunetics 3, ed. Lodewijk Bos, L. Roa, K. Yogesan, et al. IOS Press, 2006.
7. W. Hammond. Health Level 7: A protocol for the interchange of healthcare data. In Progress in Standardization, In Health Care Informatics, Georges J.E. De Moor, C. McDonald, and J.N. Van Goor, eds., Amsterdam: IOS Press, 1993.
8. A. Hutchison, M. Kaiserswerth, M. Moser and A. Schade., Electronic data interchange for health care. Communications Magazine, IEEE 34: 28–34, 1996.
9. B. Smith, and W. Ceusters. HL7 RIM: An Incoherent Standard. En Studies in Health Technology and Informatics. Ubiquity: Technologies for Better Health, In Aging Societies – Proceedings of MIE2006, A. Hasman, R. Haux, J. van der Lei, and F.H. Roger France, eds., 124:133–138. Amsterdam: IOS Press, 2006.
10. L. Schmitt, L. Schmitt, T. Falck, T. Falck, F. Wartena, and D. Simons. Novel ISO/IEEE 11073 Standards for Personal Telehealth Systems Interoperability. In High Confidence Medical Devices, Software, and Systems and Medical Device Plug-and-Play Interoperability, 2007. HCMDSS-MDPnP. Joint Workshop on, 146–148, 2007.
11. I. Martinez Ruiz, M. Galarraga, L. Serrano, P. de Toledo, S. Jiménez-Fernández, J. Escayola, E.A. Viruete, J. Fernández, and J. García. Enhanced Solutions for

Healthcare Telemonitoring in Ambient Assisted Living based on ISO/IEEE11073 standard. Methods of Information in Medicine Special Topic Issue: Smart Homes and Ambient Assisted Living in an Aging Society. pp. pte (ISSN: 0026-1270. IF2005: 1.235), 2007.

12. J. Froehlich, and J. Krumm. Route Prediction from Trip Observations, In Society of Automotive Engineers (SAE) 2008 World Congress, 2008.
13. A. Fernández-Montes, J.A. Álvarez, J.A. Ortega, N. Martínez Madrid, and R. Seepold. An Orientation Service for Dependent People Based on an Open Service Architecture, HCI and Usability for Medicine and Health Care, Springer LNCS, Volume 4799, ISSN 0302-9743, 2007.
14. D. Patterson, and L. Liao. 2004. Opportunity Knocks: A system to provide cognitive assistance with transportation services. Proceedings of Ubicomp 2004.
15. Alive Technologies, http://www.alivetec.com ,2008.
16. R. Lee, K. Chen, C. Hsiao, and C. Tseng. A Mobile Care System With Alert Mechanism. Information Technology in Biomedicine. In IEEE Transactions on 11(5), (September): 507–517. doi:10.1109/TITB.2006.888701, 2007.
17. M. Ibañez, N. Martínez Madrid, and R. Seepold. Virtualization of Residential Gateways. In Proceedings of the Fifth International Workshop on Intelligent Solutions in Embedded Systems (WISES07, ed. Ralf Seepold, Natividad Martínez Madrid and Markus Kucera, 115–126. Leganés (Spain): Universidad Carlos III de Madrid, 2007.

Chapter 10

Transparent IP Cores Integration Based on the Distributed Object Paradigm

Fernando Rincón, Jesús Barba, Francisco Moya, Félix J. Villanueva, David Villa, Julio Dondo and Juan Carlos López
University of Castilla-La Mancha, Ciudad Real, Spain, cdlujan@atc.us.es

Abstract Heterogeneous system architectures are currently the main platform on which an ever increasing number of innovative applications (i.e. smart home or ambient intelligence applications) rely. When designing these complex systems, one of the most time-consuming tasks is the definition of the communication interfaces between the different components through a number of scattered heterogeneous processing nodes. That is not only a complex task, but also very specific for a particular implementation, which may limit the flexibility of the system, and makes the solutions difficult to reuse. In this chapter, we describe how to provide a unified abstraction for both hardware and software components that have to cooperate with each other, independently of their implementation and their location. Based on this abstraction, we define a low-overhead system-wide communication architecture that offers total communication transparency between any kind of components. Since the architecture is highly compatible with standard object-oriented distributed software systems, it also enables seamless interaction with any other kind of external network.

Keywords System-on-Chip, IP CORE, Distribuited Object, HW/SW codesign

10.1 Introduction

Latest consumer applications (e.g. multimedia processing or 3D games) demand complex designs to meet their real-time requirements while respecting other design constraints, such as low-power or short time-to-market. In this

N. Martínez Madrid, R.E.D. Seepold (eds.), *Intelligent Technical Systems*, Lecture Notes in Electrical Engineering 38, DOI 978-1-4020-9823-9_10,

context, Systems-on-Chips (MPSoCs) have been proposed as a promising solution. Nevertheless, one major challenge in such systems is the integration in the platform of the multiple Application Programming Interfaces (API) that each component (e.g. memory, buses, cores, etc.) is designed for. Moreover, another important problem in SoC design is the knowledge of the position of each component in the final system to be able to efficiently communicate with it (e.g. local, remote), which makes the correct design of a SoC even more complex. Thus, new methods that allow designers to get unified inter-communication methods on SoCs architectures in the system integration flow are in great need.

Some concepts taken from distributed object platforms such as CORBA or Java RMI have already been applied to SoC design in order to get a unified view of HW and SW modules. In this paper we present an approach which inherits most of these previous achievements enriched with a strong focus on location transparency and network transparency. The resulting architecture provides a unified view of the whole system and also enables the designer to seamless develop multi-SoC systems with different network technologies.

This paper is organized as follows. In Section 10.2 we present a motivational example that will serve to guide us through our proposed approach to homogeneous hardware and software modeling. In section 3 we present the overall hardware SoC architecture of a system level middleware. In section 4 we revisit the motivational example under the distributed object approach described in Section 10.3. In Section 10.5 we show some experimental results. In Section 10.6 we overview some related work. Finally, Section 10.7, summarizes the contributions of the paper and presents possible future research directions.

10.2 Motivational Example

Let's consider the design of a System on Chip for applications with some cryptographic requirements. For such purpose, the system will include a third party DES IP core that is able to provide encryption an decryption of certain blocks of data. We consider three different usage scenarios of the DES core. In the following paragraphs we will first describe each of the situations and the problems found in typical approaches, while in the next section we will analyze an alternative solution.

For the example we will use one of the DES cores provided by Opencores [1], that will also be used to illustrate the experimental results in Section 10.5.

10.2.1 HW-HW Integration

The first scenario will be the use of the DES IP core from another hardware component. The DES core obtained from Opencores has a very simple interface with an encrypt and decrypt signal, a bus for providing the key, and the input and output data buffers.

Let's suppose that the SoC uses an OCP [2] bus for cores and processors interconnection. The first task would be to adapt it to that concrete bus. For such purpose we could use the CoreCreator tools from the OCP suite, and automatically generate the OCP wrapper out of a simple description of the DES interface. However, there is still some work to do, since we need to serialize the reception of the key and data input block, and the transmission of the data output block, since they do not fit in the bus word size.

On the other side, for the core using the DES, we should follow a similar but inverse procedure. Once known the bus interface, and the transaction-level protocol for providing the data and obtaining the results, we could write the functionality of the client core, next include some logic for the serialization of the transmission, and wrap it automatically to the OCP bus. This is a very normal procedure for IP integration, where both components are attached to the bus through some bus adapters (wrappers).

One of the advantages of using bus standards such as OCP is that they ease reusability of previous designs. However, interoperability at the level of operations is not guaranteed by the standard. The definition of a certain order of the key and data arguments is hardcoded inside the wrapper. Also the way arguments are divided into bus-size words is completely implementation dependent. That is the reason why cores with the same interface may not be interoperable.

All those problems are due to the loose coupling between functionality and communication. Therefore, any change in the component designed will also affect all of its clients, and may also imply their redesign.

10.2.2 HW-SW Integration

As a second scenario, we will consider how the DES could be used from a software client. To do so, in the classical approach we would need some kind of driver or API interface. These APIs are very dependent on the concrete core, and not easy to generate automatically. Even one minor change such as the modification of the address of the target component, or the order in which arguments are transmitted would require major modifications in the code.

Moreover, the APIs can be influenced by the specific transport architecture. For example, we could consider to add to the DES core the possibility to receive a set of words for batch processing with the same key. Transmissions from the CPU are performed word by word, and therefore can not take advantage of high-performance facilities such as bursting. Even, since a hardware client will probably transfer the block as a burst, it may be necessary to provide two different interfaces for both hardware and software clients.

10.2.3 Remote Communication

The third scenario will correspond to the request of the cyphering from outside the SoC. This may be the case for a pervasive computing application that needs some cyphering, for example. This interaction may be carried out though a wireless interface, for example.

This scenario is not very common, mainly due to the difficulty of multiplexing the ethernet between several components plus the microprocessors, which normally act as the masters of the device. Even, it is not clear how to translate network packets into the required bus transactions for the arguments and results of the operations. It would, however, be very useful to have remote communication to and from external clients, to make special computational resources accesible, for debugging purposes, for remote configuration, or even for remote reconfiguration of the SoC.

10.3 The System-Level Middleware

Most of the problems that SoC designers face nowadays are recurrent, and they have been tackled for decades in heterogeneous distributed computer networks environments. Since the 90's, the use of a system middleware has been the satisfactory solution in this field. Although it can be established a correlation between the existing problems in computer networks and SoCs, the extension to the latter is not straightforward since they have their own special requirements, such as low power consumption, or low execution overhead, for example.

A middleware is an abstraction layer whose main objective is to provide an homogeneous communication mechanism between the components of a distributed system. Generally, a middleware bases its functionality on: (a) a client-server model of communication, (b) a common data type system and a set of data coding/encoding rules, and (c) a simple protocol defining the set of messages client and server exchange. The objective is to provide orthogonalization between behavior and communication.

Applications using the middleware are usually based on the object-oriented programming model. Objects also rely on a simple communication model: method invocation. This same mechanism is used for remote communication (Remote Method Invocation or RMI), where invocations are translated into synchronous messages passed though a certain communication infrastructure. The main advantage of RMI is that it provides a neat separation between functionality and communication. That makes Distributed Object Systems specially suited to deal with heterogeneity and scalability of applications.

In RMI any method invocation must take place between certain adapters, a Proxy (the client adapter) and a Skeleton (the server adapter). From the client's point of view, the proxy is the requested object itself, since it provides exactly the same physical interface. On the other hand, server objects do not need to care about the location of client objects. They just provide an object interface which is exported through a skeleton. Thus, proxies and skeletons completely hide the real communication process. Also, in most standard software middlewares, the approach described above relies on the automatic generation of the proper proxies and skeletons depending on the kind of communication that must be established between objects.

We could consider the SoC just as another type of distributed system. Like such systems, a SoC is composed of a set of heterogeneous computing and storage resources linked through some interconnection infrastructure, and suffers the same kind of problems: scalability, heterogeneity, different communication technologies, etc. Hence, it seems reasonable to apply the same kind of solutions, and concepts, although not necessarily the same implementation.

In the following paragraphs we briefly describe the main components of the system middleware (Object Oriented Communication Engine – OOCE).

10.3.1 The Communication Broker

This layer of the middleware distributes remote invocations from the clients to the servers. In software systems it is normally a layer built on top of the operating system.

One of the main differences in OOCE with respect to the communication broker is that all components in the system share a physical communication infrastructure, which can be a bus or an on-chip network. The bus (or network) is already able to route the messages from one object to another, so there is no need of an extra layer for such purpose. Even for software objects, there is no extra layer, but remote invocation is a communication primitive. Thus there is no need of an operating system to provide remote communication.

10.3.2 Proxies and Skeletons

Proxies and skeletons provide transparency, in 3 different aspects:

1. In the location of the target, which is normally coded in the proxy, and not hardcoded in the object (the functionality)
2. In the implementation technology of the objects. SW or HW proxies will generate exactly the same transactions in the bus. That makes it impossible to know if the invocation came from a HW or SW object, as it also happens with the response.
3. In the communication technology employed. This relates to how addresses for bus transactions are built from the target object and the operation requested; how the arguments are ordered, so all requests for the same operation are always performed the same way, with independence of the source; and how data types are serialized for their transmission through the bus.

Finally, we should highlight that proxies and skeletons can be generated automatically from the object interface description, so objects can be reused under any other different context (another bus protocol, for example) just regenerating the corresponding adapters.

10.3.3 Hardware Cores

A hardware core in the SoC will be the combination of three parts: (1) the hardware object, which contains pure functionality; (2) one skeleton, as an adapter for those operations that the object is able to serve; (3) as many proxies as the object uses as a client.

From all three parts, only the object is meant to be reusable, while skeletons and proxies should be efficiently generated depending on each particular case.

But even cores not been designed with this approach in mind may be used in the system middleware. For example, any RAM memory can be seen as an object providing read and write operation for bytes, words, double words, or even larger data blocks. The only thing required is a proxy that translates such operations into the proper transactions (DMA access for a block transfer, for example).

10.3.4 CPU Adapter

The main difference between hardware and software objects (in the OOCE context) is that software objects share a common processing element, while hardware objects execute in their own. This makes it necessary some multiplexing mechanism for SW clients to have access to the bus. This multiplexer is called the Object Adapter, and consist in a set of SW routines with a standard API that must be linked with the object code of the SW clients. For every object to be able to have access to the bus, first it must be registered in an Object Adapter.

Another problem with HW to SW invocations is that objects inside the CPU are not visible out of it. CPUs are usually just masters of the bus, and are not addressable. Here the solution adopted has been to insert a bus interface between the CPU and the bus. In SW to HW invocations, the interface simply buffers the invocation and translates it into a bus transaction. In HW to SW invocations, the interface holds a translation table with bus addresses and object identities. If any of these addresses is detected in the bus, the interface buffers the transaction and notifies the Object Adapter in the CPU through an interruption. The OA then routes the invocation to the proper object, and the response back to the interface, if there is one. The interface then provides the server capabilities to the objects inside the CPU.

10.3.5 Remote Bridge

The aim of the remote bridge is to translate internal (to the SoC) invocations to external ones through some kind of network interface. The information that must be transmitted on both sides of the communication has already been serialized, so the main task of the bridge is to pack it into the messages for a certain network transport protocol.

On the SoC bus side the bridge listens for transactions addressed to external objects. Those are recognized through an internal translation table, where some internal addresses are mapped to the network addresses of the referred objects.

On the network interface it performs the opposite task. In any case, messages coming in and out of the interface have always exactly the same format as internal interactions.

10.4 The DES Example Revisited

The distributed object paradigm establishes a clear separation between the programming model and the arquitecture supporting it. Also, the OOCE platform allows the transparent integration of either hardware or software components. Thus, we can distinguish three different roles during the implementation of the system. On one side the typical hardware and software engineer roles. On the other side an integration specialist is required for the design and integration of the communication platform, as the backbone of the subsystems.

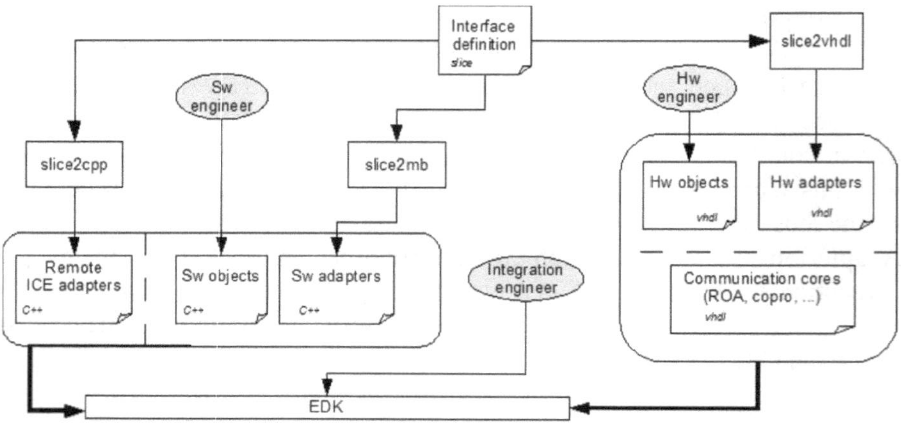

Fig. 10.1 System Design Flow

Figure 10.1 shows the relationship between the three roles, as well as the flow for the automatic generation of the architecture, that once integrated with the rest of hardware and software entities becomes the final system. This system is not limited to one chip, but can also include components deployed on other types of computation nodes linked through a communication network.

```
module slice_example {
   ["hw:bus:plb", "hw:bus:args:64"]
   interface DES {
      long int encrypt(long int key, long int data_in);
      long int decrypt(long int key, long int data_in);
   };
};
```

Fig. 10.2 Slice definition file for the DES core

The starting point of the flow is the interface specification file. This file includes the description of the interfaces for each object in the system. They are specified using an interface definition language (IDL) which is implementation neutral. Since OOCE is inspired in the ICE [3] middleware, the IDL is expressed using the Slice language. Figure 10.2 shows the slice interface definition for the DES example.

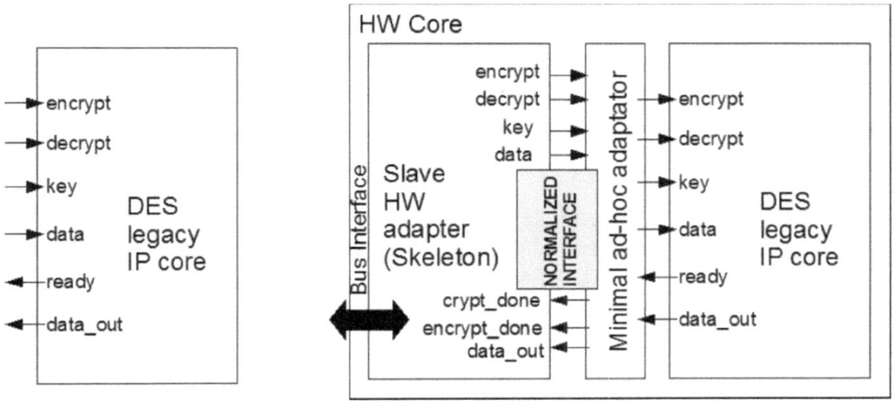

Fig. 10.3 DES Hardware Component. a) DES legacy IP core. b) DES object + adapter

Although the interface description is implementation neutral, and since the platform must be generated from this file, it may include some metadata to guide the synthesis tools. It may provide extra information such as the implementation technology (hw or sw), the bus protocol (OPB, PLB, OCP ...), the communication type (asynchronous or synchronous, blocking or non-blocking ...), etc. This is in fact one of the tasks of the system integrator, to annotate the interface definition file, generally through iterative refinement to provide the platform that best suits a certain system.

The Slice file is parsed by different code generators. Each of them, depending on the metadata, will generate specialized adapters for every object and context. The VHDL code generator (*slice2vhdl*), for example, will write synthesizable vhdl models for the adapters that will be appended to the clients or servers developed by the hardware engineers, obtaining the different cores of the system.

In case reusing cores are not designed with the distributed object approach, some extra logic is required to adapt their legacy interface to the one derived from the slice method signatures (Fig. 10.3b). The overhead for the DES core is almost negligible due to the simplicity of the normalized interface proposed. However, writing the hardware object from scratch will not incur in such overhead. The object will only include the functional code for implementing the operations, and will leave the communication responsibilities to the generated adapters.

From the same slice definition, the *slice2cpp* generator will produce the equivalent adapters (proxies and skeletons). Those adapters are based on a function library which implements the link between the CPU and the coprocessor for the communication. A device driver is no longer a collection of low-level reads and writes to a register bank interface. Now, the programmers deal with software objects that are instances of the proxies to the hardware models.

```
class DES {
  public:
    long int encrypt(long int key, long int data_in) {
      // the object identity and operation identity are mapped onto an address
      putfsl(OBJ_ID<<16 + ENC_ID);
      // the bus interface needs to know the number of arguments
      putfsl(NON_VOID|ENC_ARGS);
      // arguments and return values are serialized as two 32 bit words
      putfsl(key & 0xFFFFFFFF);
      putfsl(key >> 32);
      putfsl(data_in & 0xFFFFFFFF);
      putfsl(data_in >> 32);
      getfsl(data_out_low);
      getfsl(data_out_high);
      return data_out_high << 32 + data_out_low;
    }
}
```

Fig. 10.4 C++ code for the DES SW client

Figure 10.4 shows how the distributed object model eases the task of the software developer. Here the use of the DES core is concerned with the invocation of the methods provided in the proxy, which completely hides all the implementation details of the communication. The proxy behavior with respect to the blocking or not of the executing thread may be configured in the slice definition file, providing a high degree of control to the programmer. This is completely orthogonal to the way communication is implemented through the bus.

The final task for the system integrator is the combination of the different adapters with hardware and software objects to build the hardware cores and software components, and the inclusion of the rest of components of the OOCE engine.

To illustrate the robustness of the approach, let's suppose that the target platform is modified and the bus protocol and the bus size are now different. Those changes will not affect either the program using the DES component, or the adaptation of the DES core to the normalized interface. No modification will also be required to any of the other hardware clients (hardware components using the DES). It will only be a question of automatically regenerating the corresponding adapters.

In the proposed approach, remote communication does not imply nothing but a special bridge connected to the ethernet adapter. As it happens with SW to HW communication, interoperation is guaranteed by the use of the same bus transactions for the same operation requests. Thus, once a network packet coming from the outside reaches the bridge, it is injected in the SoC bus just if it was generated locally. The target core will recognize the address and perform the required operation putting back the results in the bus. These results are translated in the bridge back into network packages, and sent back to the remote client.

It is also possible to execute operations from remote servers from both hardware and software clients. They simply need the corresponding adapter (with the target server interface) that will translate operation requests into bus transactions. However, this transactions will not correspond to any address in the SoC address space, but will be mapped to the bridge. So the bridge will pack them into network packets, after a translation of the local SoC address to a network address (protocol and port), and will route them through the ethernet device.

10.5 Experimental Results

As a proof of concept, the OOCE has been fully prototyped on the Xilinx XUP-V2Pro platform. We have performed a set of experiments, where we have considered all the different communication mechanisms, and tested all possible interactions between components. We have also characterized the results in terms of latency (Table 10.1) and area (Table 10.2). The types of interfaces refer to: (I) simple synchronous and blocking read and write operations (without bursts), (II) simple asynchronous and non-blocking read and write operations, and (III) same as II using bursts.

Also a completely SW version of the DES algorithm was implemented on the Microblaze 32 bit processor, to use it as the software reference model. Next, all the middleware infrastructure was generated for a SoC with HW, SW and remote clients for the DES model from Opencores.

Table 10.1 Communication latency for all types of OOCE interations

Iface	Invocation type	Latency (cycles) write/read
Type I	Hw –> Hw	3 / 2
	Sw –> Hw	N.A. / 10
	Hw –> Sw	N.A. / 11
	Sw –> Sw	50 / 21
Type II	Hw –> Hw	4 / 2
	Sw –> Hw	21 / 10
	Hw –> Sw	21 / 11
	Sw –> Sw	42 / 21
Type III	Hw –> Hw	19 / 17
	Sw –> Hw	56 / 27
	Hw –> Sw	56 / 29
	Sw –> Sw	108 / 56

Table 10.2 Area cost for the hardware adapters

Interface type	Resource	Area
Simple R/W	Hw proxy	4 FFs 7 LUTs
Simple R/W	Hw skeleton	2FFs 153 LUTs
Async R/W + burst support	Hw proxy	102 FFs 208 LUTs
Async R/W + burst support	Hw skeleton	102 FFs 208 LUTs

Results measured for the DES encryption of a 2KB data block with a 55-bit key where the following: 102 microseconds for the fully SW version, 7 microseconds for the encryption using the DES core and a software client, and 5 for the completely hardware solution.

Also communication times for an off-chip invocation through an ethernet interface were measured. The reception of the packet took 218 cycles. The remote bridge translated the message into a bus transaction in 76 cycles. The execution of the invocation took 16 cycles. Finally, the result was packed into an Ethernet frame in 72 cycles and transmitted back in 218 additional cycles.

10.6 Related Work

The ideas presented in this paper complement previous work on system-level abstractions. Orthogonalization of concerns in system-level design as proposed by Keutzer [4], and more recently by Cesario [5] and Gertslauer [6], provide an object model similar to what this paper assumes, but most actual implementations focus on a structural view of the system and do not care about location transparency. In Mignolet [7] a uniform communication mechanism for HW and SW resources is proposed, based on a central HW-SW Operating System and a HW abstraction layer to provide task abstractions for HW components. Previous works by Paulin et al. [8] already apply concepts from distributed object middlewares to SoCs but they do not even consider one of the key features, location transparency. Some early ideas on how reconfigurable computing may benefit from these concepts are found in Hetch [9]. Previous results on automated generation of communication infrastructure for SoC design [10,11] are also applicable to adapters generation.

Object-based and object-oriented approaches [12,13] have also been used extensively to reduce the effort of translating some software components into hardware components or to improve the co-simulation of the system. Our hardware objects require a subset of what is provided by these extensions. Therefore we remain compatible with their approaches and we also keep full compatibility with standard IP based methodologies.

10.7 Conclusions

The communication architecture presented in this paper extends the distributed object paradigm to SoC platforms. The proxy and skeleton abstractions plus RMI semantics, provide a simple way to decouple component functionality from communication implementation. From the designer perspective, this provides an homogeneous view of the system as a collection of communicating objects. From the implementation point of view, the model presented provides communication and location transparency for any kind of local interaction between hardware and software components, blurring the hardware and software interface barrier. But it also provides the possibility of remote (may be off-chip) interaction with other objects.

Moreover, all the services and components that are part of the middleware can automatically be generated based on a few descriptions on the interfaces of the objects. This enhances the possibility of future reuse and eases design space exploration tasks. And, as the experimental results show, the communication architecture does not incur in high overheads.

Acknowledgments This work has been funded by the Spanish National and the Castilla-La Mancha Regional Goverments under grants TIN2005-08719 and PAI 08-0234-8083, respectively.

References

1. Opencores; http://www.opencores.org; last visited June, 27, 2008.
2. Open Core Protocol (OCP); http://www.ocpip.org, last visited June, 27, 2008.
3. Internet Communication Engine (ICE); http://zeroc.com, las t visited June, 27, 2008.
4. Keutzer, K., Newton, A.R., Rabaey, J.M., and Sangiovanni-Vincentelli, A. System-level design: orthogonalization of concerns and platform-based design. IEEE Trans. Computer-Aided Design of Integrated Circuits and Systems, 19, 12 (Dec. 2000).
5. W. Cesario, L. Gauthier, D. Lyonnard, G. Nicolescu, and A.A. Jerraya. Object-based hardware/software component interconnection model for interface design in system-on-a-chip circuits. The Journal of Systems and Software, 70, 2004.
6. A. Gerstlauer, D. Shin, R. Dmer, and D. D. Gajski. System-level communication modeling for network-on-chip synthesis. In Proceedings of theASP-DAC, 2004.
7. J-Y. Mignolet, V. Nollet, P. Coene, D. Verkest, S. Vernalde, and R. Lauwereins. Infrastructure for design and management of relocatable tasks in a heterogeneous reconfigurable system-on-chip. In Proceedings of the DATE '03 Conference, 2003.
8. P.G. Paulin, C. Pilkington, M. Langevin, E. Bensoudane, O. Benny, D. Lyonnard, B. Lavigueur, and D. Lo. Distributed object models for multi-processor SoC's, with application to low-power multimedia wireless systems. In Proceedings of the DATE '06 Conference, Munich, Germany, 2006.
9. R. Hecht, S. Kubish, H. Michelsen, E. Zeeb, and D. Timmermann. A distributed object system approach for dynamic reconfiguration. In Reconfigurable Architectures Workshop (RAW 06), Rhodos, Greece, April 2006.
10. V. D'silva, S. Ramesh, and A. Sowmya. Bridge over troubled wrappers: Automated interface synthesis. In Proceedings of the Intl. Conf. on VLSI Design, 2004.
11. A. Gerstlauer. Communication abstractions for system-level design and synthesis. Technical Report CECS-TR-03-30, UC Irvine, 2003.
12. Grimpe, E., and Oppenheimer, F. Extending the SystemC Synthesis Subset by Object-Oriented Features. In Proceedings of CODES+ISSS, Oct. 2003.
13. Schulz-Key, C., Winterholer, M., Schweizer, T., Kuhn, T., and Rosenstiel, W. Object-Oriented Modeling and Synthesis of SystemC Specifications. In Proceedings of theASP-DAC, 2004.

Chapter 11

Platform Modeling in Safety-Critical Embedded Systems

Bernhard Huber and Roman Obermaisser
Institute of Computer Engineering, Vienna University of Technology, Austria, huberb@vmars.tuwien.ac.at

Abstract This paper describes a model-based development process for safety-critical embedded real-time systems that are based on the DECOS integrated architecture. The DECOS architecture guides system engineers in the development of complex embedded real-time systems by providing a framework for integrating multiple application systems within a single distributed computer system. This integration is supported by a model-based development process which enables the reuse of application software on different instantiations of the DECOS platform, performing validation activities earlier in the development phase, and a reduced time-to-market in spite of increasing system functionality. For this purpose, model-based development in DECOS distinguishes between the capturing of the application functionality in a platform-independent model and the specification of the characteristics of the execution platform in the platform model. In this paper, we focus on the modeling of the execution platform and present a novel graphical model editor based on GME for specifying the DECOS execution platform. A platform meta-model expressed using UML and OCL constrains developers in such a way that the ensuing system becomes more dependable, maintainable and supports composability.

Keywords Model-based design; Integrated architectures; Embedded real-time systems

11.1 Introduction

Over the past two decades, the world of embedded systems has substantially changed. Mainly driven by competitive pressure and market forces of designers to introduce new products that set themselves apart from competitors, the functionality of embedded systems as well as the number of

N. Martínez Madrid, R.E.D. Seepold (eds.), *Intelligent Technical Systems*, Lecture Notes in Electrical Engineering 38, DOI 978-1-4020-9823-9_11,

interacting components has steadily increased. For instance in the automotive industry, the replacement of many hydraulic control systems has risen the number of micro controllers per vehicle from an average of 20 to 40-60 between the year 2000 and 2003 [1]. Thus, the key challenge faced by embedded system developers has become to manage the ever increasing complexity of such distributed real-time systems. The management of this complexity is aggravated by the inadequate abstraction level of today's programming languages. Many present-day embedded systems are *defect intolerant*, in the sense that even the smallest defects can cause major and expensive failures [2]. This observation is based on the insight that a semantic gap exists between the means of expression in programming languages and real-world problems. Therefore, it is thus suggested to raise the level of abstraction of program specification to a level that is closer to the problem domain [2].

Following this line of reasoning, the present paper describes a contribution towards a model-based development process for integrated systems. Integrated computer systems such as DECOS [3] contain a multitude of application subsystems (e.g., power-train, comfort, multimedia, safety in a car) which share a common distributed real-time system. The presented model-based development process starts with two models at a high level of abstraction, namely a platform-independent model that specifies the application services and a model of the execution platform. Using these two models, embedded system developers can specify the relevant properties at the application and platform-level in a form that is close to their problem domain. The models are used as an input for a tool-chain, which creates a platform-specific model in which the application services have been mapped to the available resources of the execution platform (e.g., computational resources of node computers, communication resources of networks).

The present paper contributes to an essential part of the model-based design process of DECOS—the modeling of the execution platform. It argues for the separation of resource modeling into two phases, which mainly facilitates re-use and hierarchical composition of platform models. Furthermore, the paper shows a prototypical implementation of a graphical editor that fits to the DECOS tool-chain. This tool is based on the Generic Modeling Environment (GME) and provides an intuitive and convenient front-end for modeling the execution platform. It expedites the modeling activities via generic templates for all constituting parts of the execution platform. Furthermore, by enforcing consistency checks and structural constraints, the majority of design faults are ruled out right from the beginning.

The chapter is structured as follows. Section 11.2 gives a short overview on related work. The model-based development process is outlined in

Section 11.3. Section 11.4 describes the devised platform model editor based on GME, which is employed for modeling an exemplary DECOS platform. The results of this case study are described in Section 11.5. The paper concludes with a discussion in Section 11.6.

11.2 Related Work

This section exemplarily describes important related work on model-based design for embedded systems and elaborates on the contributions of the proposed framework compared to those existing solutions.

11.2.1 Marte

MARTE [4] (Modeling and Analysis of Real-Time and Embedded Systems) is a UML profile that offers a unified modeling language for real-time embedded systems. MARTE was developed to enable tool interoperability and facilitate training of system and software engineers through a common language for real-time embedded systems. UML does not provide concepts for fully capturing real-time systems, such as time, resource, scheduling. Therefore, MARTE introduced time and resource models, support for modeling non functional properties, and platform modeling. In analogy to the work presented in this paper, MARTE distinguishes between logical and physical views and is compatible to the Model Driven Architecture (MDA) [5]. MARTE also addresses the allocation of applications to platforms.

The major difference to the presented work is that MARTE does not constraint an implementation concerning a specific architecture. As stated by Rioux (p.17) [6], MARTE is simply a language and "*MARTE does not tell you how to design your real-time and embedded systems*".

In contrast, the framework presented in this paper constrains developers in such a way that the ensuing system becomes more dependable, maintainable, and supports composability. Technically, this is achieved using the UML meta-model in conjunction with Object Constraint Language (OCL) constrains. For example, constraints regarding the allocation of computational resources ensure that safety-critical and non safety-critical application subsystems are assigned dedicated hardware resources. This is a key for the cost effective realization of mixed criticality systems with fault isolation and support for modular certification. Likewise, constraints for the allocation of the communication resources facilitate the correct use of the network interfaces (e.g., limitations concerning topologies).

11.2.2 SysML

SysML (Systems Modeling Language) [7] is a graphical modeling language for systems engineers based on a subset of UML2.0 with extensions such as new diagrams (e.g., parametric diagram) and modified diagrams (e.g., activity diagram). SysML supports the specification, analysis, design and validation of systems. It has been motivated by the missing standard notation and semantic of UML.

The newly introduced *requirements diagram* supports the specification of relationships between requirements using stereotypes (e.g., satisfy, derive, verify). The *use case diagrams* elaborate the interactions between external users and the system. They are expressed either from the point of view of the users or from the point of view of the system. *Block definition diagrams* describe the structure of a system in a hierarchical, tree-like fashion. In order to describe behavior, *sequence diagrams*, *state machines,* and *activity diagrams* can be used. The *parametrics diagram* is a new diagram that explains relationships between parameters (e.g., dependencies between variables).

The differences of SysML to the solution presented in this paper are similar to the ones for MARTE. SysML provides a language and does not constraint an implementation concerning a specific architecture.

11.2.3 Architecture Analysis & Design Language

The Architecture Analysis & Design Language (AADL) is an approved industry standard that has been developed under the guidance of the Society for Automotive Engineers (SAE) [8]. Its core focus is modeling and model-based analysis of real-time embedded systems. Systems are modeled in terms of components and their interactions, for which AADL distinguishes two classes of components: software components and execution platform components. Software components describe the software structure including the sequence of execution in the final system using threads, processes, subprograms, and data. The hardware of embedded systems is expressed in terms of execution platform components such as processors (execution of threads), memories (storage of code and data), busses (access among components), and devices (interaction with the environment). For specifying interactions between components AADL provides ports (data, event, and event data port), which enable the directional exchange of data or events. Moreover, specialized connectors describing the access to a common shared resource such as a bus as well as for the interaction between subprograms are defined in the AADL standard.

In contrast to other modeling languages, e.g. such as UML, AADL specifies semantics for the standardized types, components and their interactions. This way, different tools have a common interpretation of AADL models, which eases the comparability of analysis results of different tools. AADL supports model interchange and tool chaining based on a standard XML/XMI definition.

Fig. 11.1 Development Methodology in DECOS

However, as explained for the MARTE UML profile, AADL provides the means for modeling and analysis of embedded systems but does not guide the system engineer in way to design embedded systems. To our knowledge, AADL provides no mechanism to define meta-models for AADL that define what a valid model of an execution platform for a particular type of systems – such as DECOS – should look like.

11.3 Model-Based Design in Decos

The DECOS integrated architecture [3] offers a framework for the development of distributed embedded real-time systems integrating multiple application subsystems with different levels of criticality and different requirements concerning the underlying platform. The DECOS development methodology adapts the distinction between *platform-independent* and *platform-specific* viewpoints as introduced by the Model Driven Architecture (MDA) [5]. For the description of the structure of distributed computer systems, the MDA introduces *models* with various levels of detail and focus such as the Platform Independent Model (PIM) and the Platform Specific Model (PSM) – and defines their role in the design of a system.

11.3.1 Adapted MDA for DECOS

By adhering to the distinction between PIM and PSM as introduced in the MDA, we separate modeling of the application from modeling of the execution platform. As depicted in Fig. 11.1 we start with the modeling of the PIM and the construction of the platform model. The *DECOS PIM* [9] decomposes the overall system (e.g., the electronics of an entire car) into a set of nearly-independent application subsystems. Each application subsystem provides an application services that is meaningful in the given application context (e.g. in the context of an in-vehicle electronic system such subsystems are the comfort, power-train, or infotainment application subsystem). Moreover, application subsystems are further subdivided into a set of *jobs*, which interact exclusively by the exchange of messages via *virtual networks* [10]. Additionally, the DECOS PIM enables the specification of dependability and performance characteristics for each application subsystem and the establishment of links to specifications of the behavior of the individual jobs, which are expressed in DECOS using SCADE [11].

The *platform model* specifies the physical building blocks of the execution platform, the node computers, with the available communication resources (e.g. network interfaces) and computational resources (e.g. memory, processors). The formal foundation for modeling the execution platform is established by the platform meta-model [12], which is outlined in the following subsection.

Using the specifications of the PIM and the execution platform, a tool-supported transformation towards the *DECOS PSM* is initiated, which performs the mapping of the logic system structure described in the PIM to the physical system structure captured in the platform model. The PSM of an application subsystem is a refinement of the PIM where the jobs have been assigned to node computers and communication resources for virtual networks are reserved on the physical network. Besides resource constraints of node computers (e.g. the memory requirements of a job must not exceed the available memory on a particular node computer) and the physical network (e.g. the available bandwidth of the physical network determines the number of simultaneous virtual networks), the transformation of the PIM to the PSM is constrained by dependability requirements. For instance, in order to increase the system reliability by means of Triple Modular Redundancy (TMR) replicated jobs have to be assigned to independent fault-containment regions (e.g. on different node computers).

In the DECOS architecture the PSM forms the input for code generation tools, which automatically produce *executable code* for the application and the platform. For this purpose, a DECOS PIM to SCADE gateway [13] has been developed which directly imports the building blocks of the PIM into SCADE, which enables the use of the qualified code generator KCG of SCADE for generating executable code. Also, automatic code generators for the architectural services (e.g., middleware for virtual networks [10] and virtual gateways [14]) have been devised. The generated source code together with the PSM forms the input to a deployment step in which the final executables for a specific instance of the DECOS execution platform are created.

11.3.2 Execution Platform Modeling

Modeling the characteristics of the execution platform is a time-intensive and error-prone engineering task. It is therefore our objective to simplify and reduce the effort for this modeling task by developing the *Platform Model Editor (PME)*. The formal foundation of this tool is established by the platform meta-model, which primarily aims at facilitating re-use and hierarchical composition of platform models [12]. For this purpose, the modeling process is separated into two phases: the resource capturing phase and the platform composition phase.

11.3.2.1 Resource Capturing Phase

This phase addresses the specification of reusable hardware entities of an instantiation of the DECOS architecture, so-called resource primitives. Resource primitives form the lowest level of the platform description. They are the atomic hardware units of an execution platform that are identified in the modeling process. The platform meta-model defines a set of common resource primitive types, which can be used across different instantiations of the DECOS architecture. These primitive types are *Processor*, *Memory*, *Communication Interface*, *Communication Controller*, and *Connector*. For each of them, a set of *hardware properties* is predefined (e.g. clock frequency for a processor resource primitive).

Moreover, for each resource primitive further hardware properties can be specified in concrete instantiations of an execution platform. In order to support the modeling of evolving types of resource primitives, the platform meta-model provides generic model entities, whose semantics and hardware properties can be freely defined. For establishing a common interpretation of newly defined resource primitives, the concept of *technical dictionaries* [15] is applied. A technical dictionary is a reference base, containing all relevant products/parts of a particular application field, where for each product/part a detailed description and a list of its properties is given. By assigning to each generic resource primitive a unique identifier of a technical dictionary, a common interpretation of the resource primitive among different developers of the model is ensured.

11.3.2.2 Platform Composition Phase

The second phase is concerned with the composition of the entire platform model out of the previously modeled resource primitives. The output of the composition is the description of a DECOS cluster with its internal structure and the network infrastructure (e.g., the time-triggered core network between nodes, external networks like field buses, etc). A *cluster* is a distributed computer system that consists of a set of node computers interconnected by a network. A *node computer* is a self-contained computational element with its own hardware (processor, memory, communication interface, and interface to the controlled object) and software (application programs, operating system), which interacts with its environment by exchanging messages. According to the DECOS model [3], a node computer is vertically structured into two subsystem. The *safety-critical subsystem* is an encapsulated execution environment for ultra-dependable applications, while the *non safety-critical subsystem* offers an environment for those applications having less stringent dependability requirements. Within the subsystems, *connector units* control the access of jobs to the shared time-triggered core network. A third connector unit, denoted as *basic connector unit* performs the primary allocation of physical network resources, as required for the separation of safety-critical and non safety-critical subsystems of a node.

The platform composition phase is structured into three different steps:

(a) **Hardware Element Composition.** As a first step in the composition phase, individual resource primitives are composed to a larger physical hardware unit (e.g., a single board computer) that is capable of realizing (parts of) a DECOS node. In the platform meta-model, these hardware units are denoted as *hardware elements* [12]. A collection of these hardware elements form a *resource library*, which provides the building blocks for the subsequent steps in the platform composition phase.

(b) **Node Composition.** The platform meta-model constrains the composition of possibly heterogeneous hardware elements to DECOS nodes. In this *node composition step*, the available resources for the safety-critical and the non safety-critical subsystems (including the resources for the execution of jobs as well as for the execution of the architectural services realizing the connector units) are specified.

(c) **Cluster Composition.** As a final step in the platform composition phase – the *cluster composition* – the interconnection of nodes forming a cluster, which represents an instance of the DECOS architecture, is specified.

11.4 Platform Model Editor

The main motivation for a tool-based modeling environment is to keep the modeling process focused on its essential challenge – the description of the available hardware resources. The user of the tool, who usually has knowledge of the setup and internals of the execution platform, should be able to describe the essential characteristics of the platform without being forced to extensively study meta-models and tools; thus, speeding up the modeling process.

The implementation of the platform model editor (PME) is based on GME, which is a configurable, graphical framework for creating modeling environments [16]. GME has been configured with a formal modeling paradigm, which contains all the syntactic, semantic, and presentation information regarding the targeted domain. The DECOS-specific modeling paradigm is derived from the platform meta-model. According to the previously described phases of the platform modeling process, the modeling paradigm comprises three different viewpoints:

(a) **Hardware Element Viewpoint.** The hardware element viewpoint is related to the resource capturing phase and determines the modeling entities for specifying the physical building blocks of the platform. It contains separate model entities for the entire set of resource primitives as well as for hardware elements. Further on, valid compositions of those model entities are defined by using containment relationships between the hardware element and the resource primitives.

(b) **Node Viewpoint.** The node viewpoint describes the structuring of a DECOS node in safety-critical and non safety-critical subsystems. It is expressed by a containment relationship between the model elements *Node* and *ConnectorUnit* (which represents the execution environment of the DECOS architectural services) and *ApplicationComputer* (which represent the execution environment of jobs), respectively, which are associated to H*ardwareElement* model elements.

(c) **Cluster Viewpoint.** The purpose of this viewpoint is to specify the required entities for modeling the cluster setup including the association of nodes to the time-triggered core network as well as to additional physical networks (e.g., field buses).

In addition to the specification of the model entities and the associations between them, each viewpoint of the modeling paradigm comprises a set of Object Constraint Language (OCL) constraints. OCL [17] is a formal language for describing constraints on models. OCL can be used to specify invariants that must hold for the modeled system during the whole lifetime or in particular system states. Consider for instance the following simple constraint expressed in natural language: *A node consists of a safety-critical and/or a non safety-critical subsystem, but at least of one of them.* Just specifying an association with a multiplicity constraint (e.g., "0..1") from node to both types of subsystems does not correctly represent this constraint, because it is still possible to model a node without any subsystem at all. This lack of information is easily added by an OCL constraint that specifies an invariant stating that the total number of subsystems must be greater than zero and less than or equal to two. Further constraints are deployed to restrict entities depending on their actual role in the model. For instance, a connector unit instantiated as basic connector unit, requires a mandatory additional interface to the core network.

11.5 Decos Execution Platform

We have applied the PME to describe a real instance of the integrated DECOS architecture – a prototypical execution platform [18] that has been developed in the course of the DECOS project. The prototype consists of a cluster of five nodes using TTP/C [19] as time-triggered core communication network. Each node hosts a safety-critical and a non safety-critical subsystem with multiple jobs of different application subsystems. Each node of the cluster comprises three distinct single board computers for realizing the basic connector unit and the safety-critical and non safety-critical subsystems.

11.5.1 Hardware Element Viewpoint

The hardware element viewpoint is used to specify the characteristics of the hardware elements, i.e. the building blocks for the composition of the execution platform. In our prototype exist two such hardware elements: the TTTech monitoring node[a] and the Soekris Engineering net4521 board[b]. The model of the Soekris board is depicted in Fig. 11.2. It specifies only those characteristics of the single board computer that are important for the DECOS development process. Thus, it describes the processor, volatile and non volatile memory for program and data storage, as well as, communication interfaces and I/O ports. All of these entities represent resource primitives of the platform meta-model.

Figure 11.2 also exemplary shows the specification of detailed information on resource primitives. Two detailed properties of the ElanSC520 CPU are exemplified here: the clock frequency and the instructions per second. It further shows the usage of technical dictionaries. In the IEC 61360 standard[c] the property *AAF224* is defined as the clock

[a] http://www.tttech.com
[b] http://www.soekris.com
[c] Standard data element types with associated classification scheme for electric components – available at http://dom2.iec.ch/iec61360/iec/61360.nsf/

frequency for micro processors enabling an unambiguous interpretation of this property for the ElanSC520 CPU.

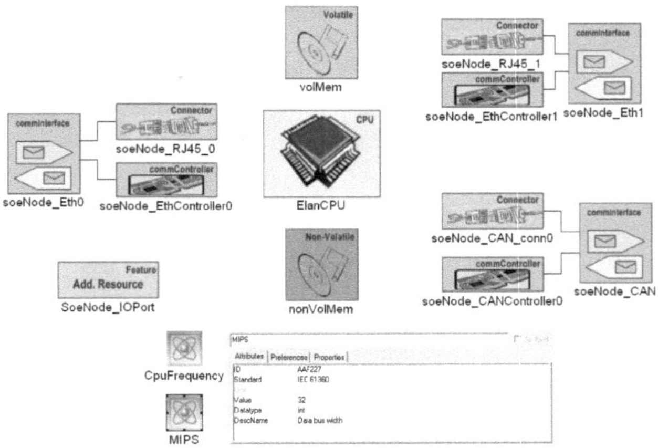

Fig. 11.2 Screenshot of models of the internal setup of soekris net4521 boards

11.5.2 Node Level Viewpoint

The node level viewpoint describes the internal configuration of a DECOS node. For the nodes of the prototype cluster we employ distinct hardware elements for the basic connector unit (cf. *MonNode0* in the left model of Fig. 11.3), the safety-critical subsystem, and the non safety-critical subsystem (i.e., one node computer for each connector unit and respective applications; cf. *SoeNode0_0* and *SoeNode0_1* in Fig. 11.3).

Those connector units are interconnected by a time-triggered Ethernet network (*ConnectorNetwork0*), which is a standard Ethernet connection using Time Division Multiple Access (TDMA) for the arbitration of the communication medium. The physical interfaces to this network are modeled by *bcuCN0*, *scuCN0*, and *xcuSC0* respectively. The basic connector unit possesses an additional interface – *CoreInterface* – with which the time-triggered core network is accessed. For the non safety-critical subsystem, an additional communication interface to a field bus network is specified.

Furthermore, it is specified whether the node provides separate encapsulated execution environments for safety-critical and for non safety-critical applications and by which hardware elements they are realized. The model on the left in Fig. 11.3 depicts a configuration in which the architectural services and the applications of one subsystem share a single hardware element.

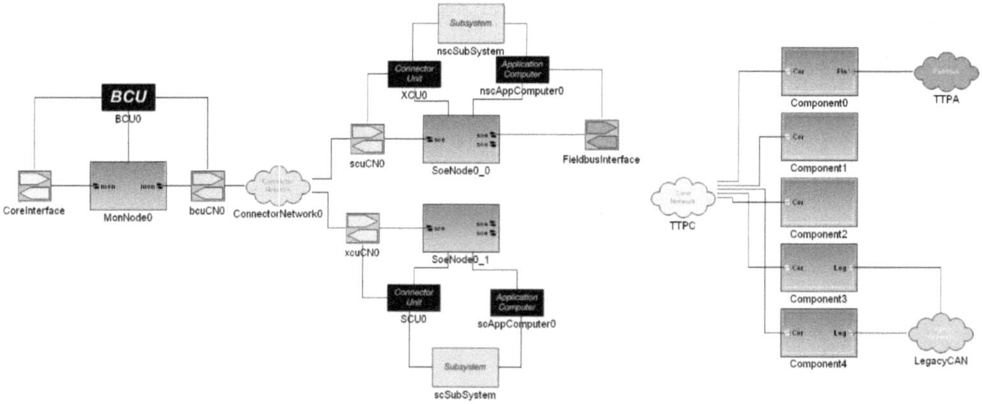

Fig. 11.3 Screenshots of node level viewpoint (*left*) and cluster level viewpoint (*right*)

11.5.3 Cluster Level Viewpoint

It is the purpose of this viewpoint to describe the nodes connected to the core network as well as properties of the core network and additional networks like field buses. The specification includes performance properties and temporal characteristics (e.g., maximum bandwidth, guaranteed latency) as well as physical characteristics like redundancy, topology, or the used physical layer. In Fig. 11.3 the cluster level viewpoint of our prototypical DECOS execution platform is depicted. It consists of five nodes, interconnected by a TTP/C network in star topology. In addition, two nodes are locally connected to a CAN [20] network while a third one accesses a LIN [21] field bus. Across all viewpoints the PME provides assistance to the user in modeling the execution platform, since the tool ensures that only models are created that comply with the DECOS modeling paradigm. Therefore, the tool prevents the creation of associations which are not specified in the modeling paradigm (e.g., in the node level viewpoint a hardware element cannot be directly connected to the network, since an interface in between is mandatory) and monitors the multiplicities of associations (e.g. it is not possible to connect communication interface resource primitives to more than one network). Finally, violations of the meta-model such as missing mandatory parts of the model (e.g., at least one subsystem has to be specified in the node) are reported to the user.

11.6 Conclusion

The development methodology of DECOS offers a model-based design process for distributed embedded real-time systems. Due to comprehensive tool support, embedded system engineers can rapidly capture the essential properties of applications and execution platforms and can perform successive model transformations down to the physical target system.

An essential part of the tool chain is the resource modeling editor. This editor enables a precise formal specification of the available communication and computational resources, while providing a user-friendly and intuitive interface to the application designer. Using a set of entities well-known in the domain of embedded systems (e.g., networks, processors, connector units, field buses), the modeling process is close to the problem domain of typical embedded system engineers. This reduces the mental effort for both, the person who creates the model and the person who is in charge of its interpretation.

Additionally, the resource modeling editor performs a pre-selection of the visible model entities to those parts that are appropriate in a given context. For instance the detailed representation of resource primitives is only visible when creating new elements of a resource library, but masked in the node or cluster viewpoint.

A further strength of the proposed modeling editor is the reduction of design faults. It is automatically checked, whether a particular relationship between two entities is permitted or not. These decisions are based on the specified DECOS modeling paradigm. Thus, violations of the underlying meta-model are directly reported to the user. Furthermore, the modeling paradigm of the execution platform model comprises various constraints specified using OCL, which further restrict entities, attributes and relationships of the modeling environment.

Finally, the presented framework supports the reuse of resource specifications through the import and export of (parts of) the platform model. Therefore, re-use of resource descriptions in different models can be simply performed by *drag and drop* operations. Moreover, the PME checks if the insertion of the model element at the specified position complies with the meta-model.

Acknowledgments This work has been supported in part by the European IST project ARTIST2 under project No. IST-004527 and the European IST project DECOS under project No. IST-511764.

References

1. C.J. Murray. Auto group seeks universal software. EE Times, 2003.
2. B. Selic. Model-driven development: its essence and opportunities. In Proc. of the 9th IEEE International Symposium on Object and Component-Oriented Real-Time Distributed Computing, page 7, April 2006.
3. R. Obermaisser, P. Peti, B. Huber, and C. El Salloum. DECOS: An integrated time-triggered architecture. E&I Journal, 3:83.95, March 2006.
4. OMG. A UML Profile for MARTE, Beta 1. OMG adopted specification. 2007.
5. OMG. Model Driven Architecture (MDA). Technical Report document number ormsc/2001-07-01, Object Management Group, July 2001.
6. L. Rioux. MARTE: A new OMG standard for Modeling and Analysis of Real-Time Embedded Systems. Thales Research & Technology, France. September 2007.
7. OMG. Systems Modeling Language (OMG SysML), V1.0 Specification., 2007.
8. SAE. Architecture Analysis & Design Language (AADL). AS5506., 2004.
9. DECOS. Dependable Embedded Components and Systems. Project deliverable D1.1.1. Report about decision on meta-model and tools for PIM specification. December 2004.
10. R. Obermaisser and B. Huber. Model-based design of the communication system in an integrated architecture. In Proc. of the 18th Intern. Conference on Parallel and Distributed Computing and Systems (PDCS 2006), pages 96–107, November 2006.
11. Esterel Technologies. SCADE Suite Technical and User Manuals, Version 5.0.1, 2005.
12. B. Huber, R. Obermaisser, and P. Peti. MDA-Based Development in the DECOS Integrated Architecture – Modeling the Hardware Platform. Proc.of the 9th IEEE International Symposium on Object and component-oriented Real-time distributed Computing (ISORC'06), April 2006.
13. W. Herzner, B. Huber, A. Balogh, and P. Csertan. The DECOS Tool-Chain: Model-Based Development of Distributed Embedded Safety-Critical Real-time Systems. DECOS/ERCIM Workshop on Dependable Embedded Systems, September 2006.
14. DECOS. Dependable Embedded Components and Systems. Project deliverable D2.2.3. Virtual communication links and gateways – Implementation of design tools and middleware services. December 2005.
15. M. Sundaram and S.S.Y. Shim. Infrastructure for B2B exchanges with RosettaNet. In Third Int. Workshop on Advanced Issues of E-Commerce and Web-Based Information Systems, WECWIS 2001, pages 110.119, 2001.
16. A. Ledeczi, M. Maroti, A. Bakay, G. Karsai, J. Garret, C. Thomason, G. Nordstrom, J. Sprinkle, and P. Volgyesi. The generic modeling environment. In Proc. of Workshop on Intelligent Signal Processing, May 2001.
17. OMG. UML 2.0 OCL specification, OMG final adopted specification. Technical Report OMG Document No. ptc/03-10-14, Object Management Group, 2003.
18. B. Huber, P. Peti, R. Obermaisser, and C. El Salloum. Using RTAI/LXRT for partitioning in a prototype implementation of the DECOS architecture. In Proc. of the Third Int. Workshop on Intelligent Solutions in Embedded Systems, May 2005.
19. H. Kopetz and G. Grünsteidl. TTP – A protocol for fault-tolerant real-time systems. Computer, 27(1):14.23, January 1994.
20. Robert Bosch Gmbh, Stuttgart, Germany. CAN Specification, Version 2.0, 1991.
21. LIN Consortium. LIN Specification Package Revision 2.0, September 2003.

Chapter 12

Service Platform for E-Safety Automotive Intelligent System

Jesús Sáez[1], Alvaro Reina[1], Ralf Seepold[1], Natividad Martínez Madrid[1], Alberto Los Santos[2], Pilar Sanz[2], Imran Sabir[3] and Henk Aarts[3]
[1]Universidad Carlos III de Madrid, Avda. de la Universidad 30, 28911 Leganés, Spain;
[2]Telefonia I+D, Emilio Vargas 6, 28043 Madrid, Spain;
[3]Philips Apptech, High Tech Campus 5, 5636 AE, Eindhoven, Netherlands,
jesus.saez@uc3m.es, natividad.martinez@uc3m.es, ralf.seepold@uc3m.es

Abstract Thousands of people die due to road accidents every year. This alarming situation has generated lots of public and private initiatives addressing the car safety improvement. This chpater aims at the definition of an intelligent system focused on taking care of the driver, passengers and vehicle conditions, so that it can be possible to prevent risky situations as well as to mitigate the damage when a critical situation has happened. To complete this target, an open service platform, characterized by a high integration level of data provided by on-board sensors and nomadic devices, has been developed, the car gateway. It allows deploying an intelligent environment for carrying out the appropriate e-safety tasks inside the vehicle. Additionally, a standalone nomadic device, the smart infant seat, has been proposed. It is in charge of taking care about the child's status. Such device will provide added-value information and functions, which will extend the e-safety routines the car gateway implements.

Keywords Automotive, Car gateway, Added-value services, Car-PC, Intelligent Environment, OSGI, Software platform, Smart infant seat, UPnP, VANET

12.1 Introduction

Several researches show that the human errors cause over 90% of the road accidents. Many of them can be prevented if the driver is aware of his physical condition and makes an effort to focus his attention or even to stop the vehicle. Furthermore, some studies have proved that the quick assistance

N. Martínez Madrid, R.E.D. Seepold (eds.), *Intelligent Technical Systems*, Lecture Notes in Electrical Engineering 38, DOI 978-1-4020-9823-9_12,

of the emergency services in a car accident scene increases the probability to save lives. Beside the automotive industrial advances, further researches are based on developing services that help us in the driving task. Conforming to this approach, the purpose of the proposal in this chapter is to develop an intelligent e-safety system able to increase the car safety and its occupant's wellness in an open automotive environment.

This system (see Fig. 12.1) is based on a modular software service platform embedded within the vehicle into a car-PC. The car gateway, which offers a set of safety applications that covers the emergency call establishment, vehicle and occupants status testing, risk situations inference and wellness guarantee. It implements an extended communication node which allows to manage the heterogeneity of the possible devices connected to the wired and wireless vehicular personal area network deployed into the vehicle environment, and offers the capability of establishing a connection with the external world (e.g. emergency services or ambulances) using cellular telephony technologies.

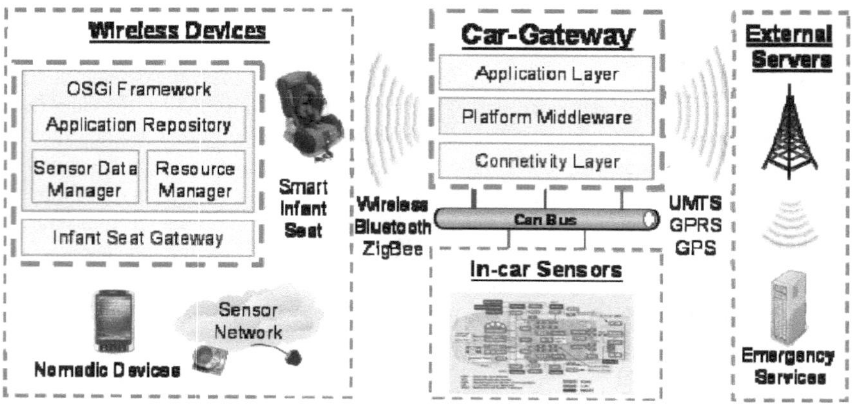

Fig. 12.1 General architecture

The developed platform implements standardized communication protocols. This feature allows supporting the integration of any new generation system that provides the final platform with added value functionality. Under the scope of this research, one of these systems, the smart infant seat, has also been developed. It can be considered as a standalone platform that improves the quality of care from parent to child using the available technology.

In the next sections, both systems are described in detail as well as their architectures and implementations. In this way, the rest of the article is organized as follows: Section 12.2 describes some of the initiatives which work on the same lines. The Sections 12.3 and 12.4 show an exhaustive description of the car gateway and smart infant seat platforms. Finally, Section 12.5 describes the currently implemented prototype and Section 12.6 summarizes and exposes the contribution of this article.

12.2 State of the Art

The i2010 program, eSafety and eCall, as well as some specific lines in the FP6 and FP7 European programs, are some of the initiatives channeled into improving the vehicular safety. Among them, several research lines are focused on developing embedded systems oriented to achieve this challenge. Regarding this kind of platforms, ADASE (Advanced Driver Assistance Systems in Europe) [1], APROSYS (Advanced Protection Systems) [2] and PReVENT (Preventive and Active Safety Applications) [3] are some of the most important integrated projects. The first one is a cluster of 30 smaller projects that works on several aspects of driver assistance, including autonomous and cooperative driving, sensors, collision avoidance, human machine interfaces and multimedia maps. By its side, APROSYS works on passive safety mechanisms, so that the technologies studied in this project are mostly based on intelligent crash and injury mitigation. Finally, PReVENT is a broad scope project, which works on several systems to detect danger situations, monitor the driver and avoid collisions. However, all these initiatives are based on non-extensible systems, which cannot accept any update based on advanced sensor and software technology.

Additionally, there exist initiatives that are in the line of the optimization of such vehicular safety platforms. Among the more interesting ones are: AIDE (Adaptive Integrated Driver-vehicle Interface) [4], EFCD (Enhanced Floating Car Data) [5] and HUMANIST (HUMAN centered design for Information Society Technologies) [6]. AIDE is an FP6 Integrated Project that addresses behavior modeling, concentrating mostly on the human machine interface for advanced driver assistance systems. As far as EFCD is concerned, it is a subproject of the Global System for Telematics (GST) Integrated Project that deals specifically with the collection of sensorial data and the communication with this data to external service centers. Finally, HUMANIST is a Network of Excellence that targets the human factor in e-safety, whose activities include cognitive driver modeling, usability analysis, user centered design and driver training and education.

To conclude, it is necessary to highlight some of the most relevant projects regarding the vehicle communications issue. Among them, there are some initiatives that are aimed at improving the integration of sensorial and control systems in the car, as iEASIS (Electronic Architecture and System Engineering for Integrated Safety Systems) [7]. In the same way, there also exist several initiatives whose objective is to promote the research of issues related to communications among vehicles, as for example MARTA (Movilidad y Automoción en Redes de Transporte Avanzadas) [8], CVIS (Cooperative Vehicle-Infrastructure Systems) [9] and Diamant (Dynamic Information and Applications for securing Mobility using Adaptive Networks and Telematics Infrastructure).

The initiative proposed within this chapter develops a multimodal system whose design covers issues related to the vehicle communications, critical situations inference, damage mitigation after an accident occurs, as well as the user-friendly features of the human-machine interface. Contrary to the other similar initiatives, the main deployment guidelines in this system are the scalability and compatibility with both current and future nomadic devices. Furthermore, this open system is based on a dynamic modular architecture that allows it to be completed by additional software modules.

12.3 Car Gateway

The car gateway is the vehicular embedded unit intended to support some intelligent applications. This unit is focused on the next two tasks: to warn the emergency services when a critical situation has occurred and to improve the wellness of the car occupants. The underlying software platform is based on the Open Services Gateway initiative (OSGi) [10]. This framework implements a dynamic component model, where any component (bundles) can be developed individually and easily integrated into the platform. So, the car gateway architecture has been designed in the OSGi platform context dividing it into three complementary layers (see Fig. 12.2).

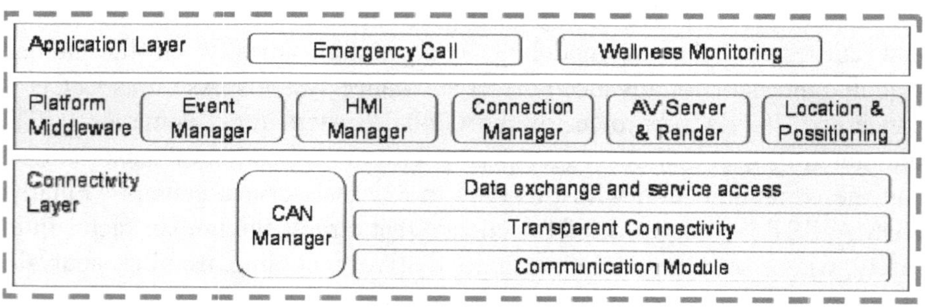

Fig. 12.2 Car Gateway Architecture

Bottom-up, a first layer (*Connectivity Layer*) is in charge of implementing and providing the services which allows the upper levels to abstract themselves from any communication routines. Over this, a middleware layer provides a set of high complexity services that support the upper layers. Finally, intelligent functionality resides at the top layer (*Application Layer*).

In the next sections this architecture is explained in more detail. Therefore, Section 12.3.1 is focused on explaining the connectivity layer. Section 12.3.2 exposes the architecture and functionality of the middleware layer. At last, the top layer is described in Section 12.3.3, defining the applications that compose the intelligent in-car environment.

12.3.1 Connectivity Architecture

This layer is focused on solving both local and remote connectivity requirements. Since the communication engine needs to be flexible and the environment is no longer static, this layer must solve the dynamic configuration, the mobility adaptation, the automatic service discovery mechanism, the compatibility with the devices heterogeneity and some security issues. It is aimed to provide an abstraction layer for vehicle applications, managing communication for multiple concurrent sessions, and spanning all communications modes and all methods of transmission.

This layer architecture is based on four modules (see Fig. 12.2) that are detailed in the next subsections.

12.3.1.1 Communication Module

This module defines the communication infrastructure of the car gateway. It constitutes an access point towards the external and personal area networks, supporting both wired and wireless technologies.

Depending on the device typology and localization, the communications can be grouped into three categories: in-car communications, external communications and sensor communications. Each one of these categories is going to be managed by one of the dedicated sub-modules which compose this entity. These are:

- **In-Car Communication Manager:** Inside the vehicle, a personal area network is deployed to communicate the internal wireless devices (nomadic devices, smart infant seat, etc.) with the car gateway. This network supports Bluetooth, WiFi and Zigbee technologies, deploying different access points suitable for such communication technology.
- **External Communication Manager:** The goal is to allow seamless mobility of data between the car and external nodes. To solve the short range communications an 802.11b/g access point, configured in ad-hoc mode, has been developed. It allows to create a VANET (Vehicular

Ad-hoc Network) among nearby vehicles and between vehicles and roadside equipment, using the AODV (Ad-hoc On-demand Distance Vector) [11] technology as routing algorithm.

The long range communications will be provided by means of GPRS/UMTS adaptors, managed by this component. Both interfaces will be working concurrently but the *Transparent Connectivity* module will decide.

- **Sensor Manager:** This sub-module acts as a new wireless sensors concentrator. It is in charge of defining the interfaces towards any sensor, using the IEEE 1451 smart transducer standards [12]. To do this, it implements the IEEE 1451.5 standard in order to deploy an access point that allows establishing the link towards the transducers. This specification covers the 802.15.1 (Bluetooth) and 802.15.4 (ZigBee) wireless communication protocol.

  On the other hand, it uses the IEEE 1451.0 standard in order to provide the upper layers with a uniform set of commands used to access any sensors in the 1451-based networks.

12.3.1.2 Transparent Connectivity Module

This module is in charge of managing the communication with any device located inside or outside of the vehicle. It is based on the TCP/IP stack, and the networking process adaptation to the vehicular environment. This module tries to adopt the CALM (*Continuous Air Interface for Long and Medium range)* [13] proposal, taking its standardized architecture as a reference. Therefore, its architecture is based on the next three sub-modules: *Interface selection, Addressing* and *Negotiation*.

Firstly, the *Interface selection* sub-module is in charge of monitoring the quality of all available links to a given destination at any moment. Based on this monitoring, and taking into account the service requirements, this system decides dynamically which link is the best to use and routes the information through it, following the 'Always Best Connected' philosophy.

Secondly, the *Addressing* sub-module is in charge of detecting any new device that wants to register into the car gateway, authenticating it, addressing it (DHCP functionality) and performing the registration at the network level. An AAA (authentication, authorization and accounting) protocol is employed for this task, currently using AAA servers that communicate using the RADIUS protocol.

Finally, the *Negotiation* sub-module provides high-level interfaces for the applications to request and setup connections with certain constraints (delay, bandwidth, etc).

12.3.1.3 Data Exchange and Service Access Module

The system consists of nomadic devices connected to the car gateway that can be disconnected and connected again by the user without prior notice (e.g. infant seat). To achieve this goal, UPnP [14] technology has been used, providing the car gateway the capacity to discover and configure automatically every device connected to the car. This protocol has been extended because the local nature of this technology, mainly due to the broadcast messages sent in the auto-discovery stage. To do this, a 'Search' routine has been included which enables to discover devices located into an external network, sending unicast SSDP discovery messages routed to its well-known IP address.

This module architecture has been implemented by a cooperative architecture of OSGi bundles, whose core is constituted by the extended UPnP protocol stack. The OSGi bundles are categorized into three branches: platform services, conversion engine, and peripheral services.

The *platform services* branch is based on two bundles that implement both an *UPnP virtual device* and an *UPnP control point.*

The *conversion engine* branch is composed by two bundles, the *Device and service access* and the *Event and notification manager.* The first one is in charge of the UPnP Service to OSGi Services transformation. The *Event and notification manager* bundle is in charge of managing the UPnP events registered by the UPnP client, translating them into java events, so it can be possible to notify it to any bundle installed over the OSGI framework.

The *peripheral services* branch is composed of a set of bundles which extend the UPnP client functionality. Among them the *AV subsystem* bundle, based on the UPnP AV Architecture [15]; it provides capabilities to negotiate AV connections between multimedia devices. Through the interface it may initiate a multimedia connection. The resources reservation for high quality connections is the goal of the *QoS Subsystem* bundle, which is based on the UPnP QoS Architecture [16].

12.3.1.4 CAN Bus Communication Module

This component acts as listener module into the car gateway, retrieving information from the Body B-CAN (Dual Wire Fault Tolerant low speed CAN), where the messages related to the comfort/safety management is transmitted. Due to security and confidentiality reasons, this module only acts in a read mode, making it impossible to introduce any message into it.

This module is formed by a stack of three specialized sub-modules. At the bottom, the *CAN interface* which is based on the ISO 11898-5 specification. It defines the physical layer of this module that provides the car gateway with connection towards the B-CAN.

Over it, the *CAN driver* provides a hardware abstraction layer for accessing the data transmitted via the B-CAN. This is based on the commercial driver obtained from the selected hardware solution provider. This driver solves the listening ports configuration, CAN messages reading and errors treatment.

Finally, at the top of the stack, the *Listener* module is installed. It is in charge of configuring the CAN port, managing any arriving message, and generating the corresponding events. Furthermore, it publishes an OSGi interface so that any upper module can access the CAN sensors data.

12.3.2 Platform Middleware

The car gateway middleware is designed as an open, modular, reusable architecture, which settles its bases in the OSGi framework. On this middleware the infrastructure supporting the e-safety applications is deployed, defining a set of high-level services (see Fig. 12.2). Among this middleware services, developed and offered to the application layer, are:

- **Event Manager:** This module registers, classifies and dispatches asynchronous Java events, acting as a notifications concentrator.
- **HMI Manager:** This component provides the applications with the presentation level and the access to the input data. The physical devices of the HMI consist of one touch screen (input-output channel), one microphone (input channel) and some speakers (output channel).
- **Connection Manager:** This OSGi bundle is responsible of managing some basic functions provided by the GSM/GPRS phone device attached to the car gateway.
- **AV Server & Render:** This module implements a Media Renderer and a Media Server according to the UPnP AV Architecture. These modules will deploy the terminal components of a videoconference or multimedia component transmission.
- **Location and Positioning:** This bundle implements a GPS/Galileo device driver. Through its OSGi interface the position (longitude and latitude), altitude (meters) and speed (km/h) can be retrieved if a GPS device is connected to the car gateway.

12.3.3 Application Layer

This layer, located at the top of the architecture (see Fig. 12.2), provides the car with the applications that watches over the car occupants safety (*Emergency Call*) and wellness (*Wellness Monitoring*). Such applications deploy the intelligent environment in charge of improving the security inside the vehicle.

12.3.3.1 Emergency Call

This application is in charge of carrying out both an automatic detection of an accident and the enriched emergency call towards the emergency services. Depending on the available communication resources, either a normal voice communication or a videoconference to the emergency center will be established. Additionally, this call is enriched by a data connection towards the emergency services that will send added value information. Among the enriched data transmitted are: vehicle identifier, time, location, personal sensors data and vehicular sensors data. Armed with such crucial data, rescue services can reach the accident scene quickly and be well prepared for the situation.

To achieve these challenges, this application is based on three bundles. Firstly, the *Environment Monitoring* is in charge of retrieving the raw data provided by the sensor network and the peripheral devices. It also supplies the other components with this information. To do this, this bundle is subscribed to the *Event manager* service from the middleware, receiving all this information of the vehicle status and its occupants in an asynchronous way. Over this module, the *Data Enrichment Engine* bundle infers when a critical situation has occurred based on the information retrieved from the environment monitoring.

At the top, the *ECall* bundle is in charge of establishing either the phone call or the videoconference with the emergency services as well as transmitting them the enriched information about what is happening in the car. This module is based on the inference process made by the previous module in order to launch the emergency call. Additionally, it uses the services provided by the *Connection manager* and the *AV server & render* modules from the middleware in order to complete this task.

12.3.3.2 Wellness Monitoring

This application monitors the driving conditions taking measurements from both the vehicle occupants and environment, in order to warn the driver and validate that the car configuration is suitable for safe and comfortable driving.

This application is implemented by a modular architecture defined by two bundles. Firstly, at the bottom, the *Wellness Inference Engine* that is in charge of reasoning whether the driver conditions are suitable to drive or not. To retrieve all the data needed to make this reasoning, it is registered by the *Event Manager* in order to receive the data of all the wellness sensors. With all the information retrieved, it can infer the wellness state of the driver, so that whenever this service detects that the driver's state is not suitable for driving, the service triggers the *Wellness Recommendation Engine* module to communicate it with the user.

The *Wellness Recommendation Engine* bundle defines the top level of this application. This engine continuously receives the current state from the previous module and is triggered by it when the driver's state is not appropriate. In the same way, it will use de services offered by the *HMI manager* in order to advise the driver about its state and give him a recommendation that could help the driver to take care of himself.

12.4 Smart Infant Seat

In the scope of this chapter, the platform is considered as a nomadic device of a new generation, formed by a sensor network located in the baby seat, and which is in charge of sensing and informing about the child's health at any moment. This disposes of communication capabilities which allow it to connect to the car gateway, providing this one with added value information that it uses in order to extend its functionality. This cooperation will enable the driver to be aware of the child status for means of the HMI system, as well as providing enriched information for the car gateway to extend both wellness and emergency call applications.

The platform architecture consists of multiple modules (see Fig. 12.1), each one specialized in a specific responsibility. At the bottom, a complete sub-system called *Infant Seat Gateway* has been developed in charge of managing all this platform communication.

Over the previous communication layer, this system deploys a middleware framework based on the OSGi specification. Simultaneously, the UPnP specification has been implemented in this device, in order to provide it with the plug and play characteristics of this technology. Inside this framework three modules in charge of implementing the complete functionality of this device are deployed. These are: *Resource Manager Module, Sensor Data Manager and Application Repository*. All of them, together with the *Infant Seat Gateway* are described in the next subsections.

12.4.1 Infant Seat Gateway

It is equipped with short range RF interfaces (Wi-Fi, Bluetooth and ZigBee) as well as wired ones (USB) which allow building ad-hoc networks with car-PC or any other nomadic devices. While Wi-Fi connection has been chosen primarily for the home environment, Bluetooth technology has been selected to implement the communications within the vehicle environment. IP support added to the Bluetooth connection as many (AV) communication and services mechanisms nowadays are based upon IP-connection.

As far as ZigBee and USB are concerned, they are used to communicate the seat gateway with the sensors deployed into its environment. Such sensors are hardware nodes specifically deployed for monitoring determinate actions (Buckle sensor, AV camera) or the child status (ECG Sensor).

12.4.2 Resource Manager Module

The role of the context manager is to acquire information from different sources such as sensors and user activities (inputs), to subsequently combine the information to predict about the context, and adapt the system to such context. Every context is related to certain situation where the services appear and disappear (enabling or disabling bundles), and interacts in a loosely and ad-hoc fashion. Furthermore, The power consumption will be managed by this module, adapting it to each situation, so that once there is no communication or no baby in the seat, it just goes into a stand-by mode and also informs the car gateway of its state.

12.4.3 Sensor Data Manager

This is responsible of collecting the raw data from sensors and if required, to apply data mining process to take out relevant information. This mining process monitors the sensor reading to be within the required threshold/frequency etc. If sensor data modules find values beyond threshold, it notifies its corresponding OSGi bundle to take further actions to notify the user. Initially, OSGi uses the PULL mechanism to activate the sensors. After that it becomes the PUSH mechanism so that data coming from the sensors is interpreted and in case of a volatile situation, it notifies its OSGi bundle to take further actions. Using the PUSH mechanism saves valuable power in mobile devices.

12.4.4 Application Repository

At the top of the architecture, an application layer is deployed. The software bundles that compose this layer implement the services in charge of acting when the previous module demands it. Also a more complex application is developed so that the child and seat status can be inferred using the raw data provided by such module.

Furthermore, these same modules define the UPnP interface oriented to provide any UPnP client (e.g. car gateway) with the processed information. Such interface will be characterized by a synchronous communication mechanism, based on UPnP services that allow the UPnP client to do the child status requests on demand, and an asynchronous communication mechanism based on UPnP events in charge of inform about this status when a relevant situation is happening. Among the information transmitted by these events are both informing messages (e.g. child and seat detection, child vital parameters) and warning messages (e.g. seat is not well restrained).

12.5 Prototype

To ensure the vehicular model works correctly a prototype of the exposed system has been developed. Such prototype includes both the service platform embedded into the vehicle and the smart infant seat as an example of a nomadic device. In this section, the details of the resources needed for such prototypes are going to be analyzed.

The on-board unit prototype consists of an extensible and flexible car-PC set that supports the car gateway deployment. This set will contain a central processing unit, the HMI interfaces and the communication interfaces. The central computing unit is composed of a motherboard (VIA-EPIA CN-1300 mini-ITX with 1.0 GHz VIA C7 processor, 1 GB DDR2 RAM and 80 GB hard disk), a rugged aluminum Voom-PC2 enclosure, and a M2-ATX Smart Automotive power supply (140W). This system is going to be embedded within the car. Finally, to develop the HMI system, it has been included audio and video (MPEG-2) hardware support for the GUI and a voice interface as well as a touch screen monitor (7" Xenarc), a microphone and several speakers and a microphone for the voice HMI.

Regarding the smart infant seat, the initial demonstrator platform has been chosen based on low-power characteristics as well as early progress. For this a low-power micro laptop has been chosen.

On the subject of the software components, a Linux kernel version 2.6 runs on both platforms. In the same way, the BlueZ PAN module [17] has been enabled for getting IP over Bluetooth compliance. On the OS, JDK 1.6

has been installed. This JVM supports the Equinox 3.2.2 framework which is the OSGi implementation selected for both prototypes. This distribution is a free implementation of the OSGi specification release 4.

The technology underlying the platform's storage mechanisms is the SQLite version 3.5.6. This database manager balances a satisfactory throughput and a small footprint. This will make it easier to embed the system into any resource-limited device. Regarding the UPnP technology, the UPnP protocol stack developed by Cyberlink [18] release 1.7 has been used. This stack has been properly modified according to platform requirements in order to solve the registration of remote UPnP services.

12.6 Conclusion

The main goal of this chapter is to describe and embedded system that will increase car safety enabling wellness and extended emergency call applications in an automotive environment. To achieve this, an open automotive infrastructure has been defined that will be constituted by a centralized system called car gateway in cooperation with sensors and nomadic devices personal area networks.

The platform proposed defines an embedded open and scalable system that allows registering and configuring in an automatic way any kind of sensors and UPnP devices. Furthermore, its software architecture, based on a modular system, allows extending the final system functioning by adding additional functional modules, or by updating the existing ones.

Among the scope of this article a new generation nomadic device has been also developed, the smart infant seat. It has been implemented as a plug and play device in charge of taking care of the child's status at any moment, informing about it to the car gateway.

The combination of both have enabled the implementation of a complete system able to deploy an intelligent environment, focused on managing critical situations inside a vehicle and helping to prevent from danger situations and improve the wellness of its occupants.

Acknowledgments The platform under study in this chapter has been partly developed inside the MEDEA+ CARING CARS [19] project, financed by the Spanish Ministry of Industry (FIT-330215-2007-1) and the InCare project, financed by the Spanish Ministry of Education and Science (TSI2006-13390-C02-01).

References

1. Alkim, T., G. Bootsma, E. Berghout, G. Ostyn, and P. Gendre, "ADASE 2 Expert Workshop on effects of ADA systems on safety, throughput and comfort", ADASE 2 Deliverable D8E, contract number IST-2000-28010, 1 July 2004.
2. IP APROSYS Sub-Project 6, available at : http://www.aprosys.com/
3. M. Schulze, G. Nocker, and K. Bohm, "PReVENT: A European program to improve active safety", IEEE Transactions on Intelligent Transportation Systems Telecommunications, pp. 322–331, ISSN: 1524-9050, 2005.
4. A. Amditis, L. Andreone, A. Polychronopoulos, and J. Engstrom, "Design and development of an adaptive integrated driver-vehicle interface: Overview of the AIDE project", Proceedings of the IFAC conference, Prague, July 2005.
5. S. Messelodi, C. M. Modena, M. Zanin, F. De Natale, F. Granelli, E. Betterle, and A. Guarise, "Intelligent extended floating car data collection", Expert Systems with Applications , ISSN: 0957-4174, Elsevier, 2008.
6. A. Pauzié, "Network of Excellence HUMANIST – Human Centred Design for Information Society Technologies INRETS & ERT", Advanced Microsystems for Automotive Applications 2006.
7. EASIS Project, available at : http://www.easis-online.org/
8. MARTA project, available at: http://www.cenitmarta.org
9. E. Koenders and J. Vreeswijk, "Cooperative Infrastructure", IEEE Intelligent Vehicles Symposium, ISSN: 1931-0587, 2008.
10. OSGi Alliance, "OSGiTM – the dynamic module system for JavaTM", URL: www.osgi.org, September 2008.
11. AODV, Ad hoc On-Demand Distance Vector (AODV) Routing, Internet Engineering Task Force (IETF), Network Working Group RFC: 3561, 2003, URL: http://www.ietf.org/rfc/rfc3561.txt
12. K. Lee, "IEEE 1451: A standard in support of smart transducer networking", IEEE Instrumentation and Measurement Technology Conference, Vol. 2, pp. 523–28, 2000.
13. ISO TC 204 Working Group, "CALM project," URL: www.calm.hu, September 2008.
14. Universal Plug and Play (UPnP) Forum, URL: www.upnp.org, September 2008.
15. Universal Plug and Play (UPnP) UPnP AV Architecture V1.0, URL: www.upnp.org/specs/av/UPnP-av-AVArchitecture-v1-20020622.pdf, June 2002.
16. Universal Plug and Play (UPnP) UPnP QoS Architecture V2.0, URL: http://www.upnp.org/specs/qos/UPnP-qos-Architecture-v2-20061016.pdf, June 2002
17. BlueZ, Official Linux Bluetooth protocol stack, URL: http://www.bluez.org, September 2008.
18. S. Konno, Tokyo, Japan, URL: www.cybergarage.org, September 2008.
19. Caring Cars, MEDEA+ project 2A-403, available at: http://www.medeaplus.org

Part IV

Sensor Networks and Autonomous Systems

Chapter 13

Intelligent, Fault Adaptive Control of Autonomous Systems

Willibald Krenn and Franz Wotawa
Institute for Software Technology, Graz University of Technology, Inffeldgasse 16B/II, 8010 Graz, Austria, wkrenn@ist.tugraz.at

Abstract We present a methodology for intelligent control of an autonomous and resource constrained embedded system. Geared towards mastering permanent and transient faults by dynamic reconfiguration, our approach uses rules for describing device functionality, valid environmental interactions, and goals the system has to reach. Besides rules, we use functions that characterize a goal's activity. These functions control the frequency our system uses to reach the goal. In this chapter we present the system model, discuss properties of the rule selection mechanism, and show results gained from running the approach on an embedded device.

Keywords Rules; Weight model; Activity factor; Damping factor

13.1 Introduction

Autonomous systems always have to deal with the unexpected. Due to unforeseen environmental interactions and/or internal faults autonomously acting systems need to know about different ways to reach a specific goal. The problem of the on-board control algorithm then is to figure out *which* way to choose. The presented approach uses information gathered by observing the system's actions in order to answer this question.

Somewhat in contrast to the emphasized redundancy, devices designed for autonomous and mobile use are often optimized for low energy consumption. A side effect of this optimization is low processing power, little on-board memory, and little redundant functionality. While these

N. Martínez Madrid, R.E.D. Seepold (eds.), *Intelligent Technical Systems*, Lecture Notes in Electrical Engineering 38, DOI 978-1-4020-9823-9_13,

conditions limit intelligent control, the presented approach was designed with these limitations in mind.

In our approach all possible ways of reaching goals are encoded in a rule set that also carries additional – developer supplied – information about the intended selection frequency ("desirability") of rules. Using this information and other factors, that are learned during executions of rules, we compute one weight for each rule. Based on this weight, our algorithm explores the search space for ways to reach goals. Similar to the presented approach, Saffiotti et al. [5] use multi-valued logic to control intelligent agents and decide based on weights between alternatives: Degrees of truth of formulae describing contextual (environmental) conditions are used to weight the preferences of different coordinated activities, e.g., following a corridor. The authors also introduce behaviors and goals and "relate behaviors to goals by defining the notion of goodness of a behavior for a goal". Behaviors are coordinated activities combined with contextual conditions and object descriptors. They can be combined and, thus, standard planning techniques can be used to generate combinations of behaviors to satisfy given goals. The biggest difference to our approach is that we create a behavior (seen from outside) by combining the effects of running all stored (independent) goals from the rule set. Furthermore, in difference to desirability functions of Saffiotti et al. [5], our "desirability" is given by the system developer and only switched in special cases. Having said that, the final weight of a rule in our approach is influenced by experiences learned in past. This can be seen to some extend as a measure for desirability as in the work of Saffiotti et al. [5] but the difference lies in the fact that we do not care about the current system state but infer this factor from the past.

Based on the idea of combining simple reactive behaviors, which we also draw upon, different versions of behavioral systems have been published: Among them are the subsumption architecture from Brooks [1], and the command fusion architecture from Rosenblatt and Payton [4]. Our approach is best compared to TR-programs [3]: The biggest difference of our approach to TR-programs lies in action selection and real-time capability. We allow dynamic updates of the rule's selection frequency, rely on discrete time steps, and do not guarantee real-time capabilities. TR-programs (consisting of TR-sequences) have a fixed priority order and real-time, circuit semantics. TR-programs also feature durative actions while our approach is based on finite actions that we use in order to simplify repair. TR-programs always evaluate all conditions. This is different to our approach, as we have discrete evaluation points. It is possible to hierarchically compose TR-sequences.

We use the example of a mobile remote sensing platform throughout the chapter for illustration purposes: The device consists of a GPS receiver, a

GSM modem, a NF module which provides medium range inter-device communication capabilities, and other hardware mostly dedicated to energy management. The device has to provide tracking services: It has to acquire the current geographical position and send it to some receiving station even if transient or permanent faults are present. Furthermore we assume that the developer has specified a preferred behavior of the system and that it only has a limited set of observations of the environment.

The remainder of this chapter is organized as follows. In the next section we present the main idea. In principle, we take a rule-based system, introduce the notion of goals, assign to each goal a function that describes how often it should be reached (max-value of function) and how important it is to reach the goal (slope of function), a damping factor that covers lessons learned from the past, and an activity factor that indicates how often a goal has been reached. After discussing further details of rule-weight calculation, we present a case study and conclude the presented work.

13.2 Basic Methodology

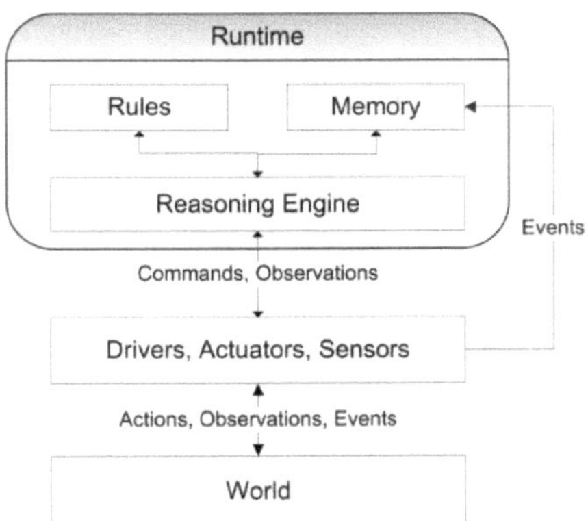

Fig. 13.1 Architecture and layering

In this section, we present the basic idea and the operational semantics that is behind our approach. Figure 13.1 shows the main parts of our control system. The rule execution engine, also called *Runtime*, is responsible for controlling the device. It comprises a rule set, main memory that stores the world state (a set of propositions), and a reasoning engine that is responsible for rule selection and world state update. External to the runtime are the driver/actuator layer and the world. We do not characterize any of the external layers any further but concentrate on the runtime instead.

We present a refined version of our system model [6] that is more explicit and has an easier to understand semantic meaning. Basis to our model are propositions that hold the current believe state of the system. Observations of the outside world are mapped by setting propositions true or false at fixed times. Besides these propositions, rules are a central element to our algorithm: Rules interconnect different propositions and also describe valid action sequences the system can trigger in order to move the system to a valid goal state. Goals, thereby, are simply specially treated rules. Whenever there are several rules that "turn on" a proposition, we use a weight function to determine the precedence.

13.2.1 Rules

We need to give a few definitions in order to present the semantics of the proposed rule-based system. We start by defining the sets the system operates on:

- Propositions P
- Memory $M \subseteq P$
- Labels L
- Rules R
- Actions A

P contains all possible propositions. If a proposition is believed to hold in the current world state then it is element of the memory, hence M is a subset of (or equal to) P. Next, we give a definition of a rule.

Definition (Rule). We define a rule as n-tuple:

$$\text{Rule } r = (\textit{label} : l \in L,\ \textit{guard} : \bigwedge_{p_i} p_i \in P,\ \textit{action} : \bigwedge_{a_i} a_i \in A,$$

$$\textit{postcond} : \{ \bigwedge_{op_1(p_i)} p_i \in P,\ p_i \text{ occur only once}, op_1 \in Ops_1$$

$$\bigwedge_{op_2(l_i)} l_i \in L, op_2 \in Ops_2 \},$$

$$\textit{activity factor} : \textit{activity} \in [0,1],$$

$$\textit{damping factor} : \textit{damping} \in [0,1],\ \textit{weight} : w \in [0,1],$$

$$\textit{max} : m \in [0,1],\ \textit{profile} : \alpha : [0,1] \mapsto [0,1])$$

Note that we restrict ourselves to post-conditions that only have one *add* operation as defined below. A rule may have assigned external actions ($\in A$) that are treated as special predicates. If the guard holds and there is no failure running the actions, the system guarantees the state described by the post-condition. The activity factor ($activity$) is a number representing how frequently the rule was chosen. The damping factor ($damping$) indicates how frequently action predicates failed, when the guard was satisfied. Note that this situation should never happen. If it does, the runtime system will run a repair action that has to report success. The activity profile function α is used for mapping the activity value $activity$ of a rule to some value in the interval [0,1]. The max value may be obtained from α: In that case it represents the input value which causes α to return a maximum value.

Definition (Operations). op_1, op_2 denote an operation from the following sets:

$$\begin{aligned} Ops_1 &= \{add(p \in P) : M \mapsto M, \\ &\quad remove(p \in P) : M \mapsto M\} \\ Ops_2 &= \{incweight(l \in L) : [0,1] \mapsto [0,1] \\ &\quad decweight(l \in L) : [0,1] \mapsto [0,1]\} \end{aligned}$$

Definition (Activity Maximum). As already said, max can be calculated from α, in which case we introduce a function

$$Max : ([0,1] \mapsto [0,1]) \mapsto [0,1]$$

that takes α and returns the input value where α becomes maximal. We need max for weight calculation, as this can be seen in the next definition.

Definition (Weight Function). Each rule gets assigned a weight which is calculated by some function

$$\gamma : [0,1] \times [0,1] \times [0,1] \times ([0,1] \mapsto [0,1]) \mapsto [0,1]$$

that calculates the weight as follows:

$$\begin{aligned} &\gamma(activity, damping, max, \alpha) = \\ &\quad \alpha'(activity) \cdot Abs(max - activity) \cdot (1 - damping) \end{aligned}$$

We assume $max = Max(\alpha)$. Before we discuss the weight model in detail, we state the definition of the interpretation of a rule.

Definition (Interpretation). A rule r is run by some function $I : M \times R \mapsto M$ defined as $I(M, r \in R) = M'$ with

$$M' = \begin{cases} M & \text{if} guard(r) \not\subseteq M \\ M \cup \{p \mid add(p) \in postcond(r)\} \setminus & \text{if} guard(r) \subseteq M \\ \{p \mid remove(p) \in postcond(r)\} & \wedge\, action(r) \end{cases}$$

and $weight(r)' = \gamma(activity(r), damping(r), max, profile(r))$
and $weight(x)' = weight(x) + c, \forall x \in R, l \in L:$
$(label(x) == l \wedge incweight(l) \in postcond(r))$
and $weight(x)' = weight(x) - c, \forall x \in R, l \in L:$
$(label(x) == l \wedge decweight(l) \in postcond(r))$

Note that activity and damping factors, as outlined in the next section, get updated by the system too. The constant c stands for a user defined increment/decrement step. When an action fails, M depends on the repair action function not discussed here and commonly provided by the developer. It is reasonable to assume $M' = M$ for the time being.

The power set of all rules $Paths = 2^{Rules}$, if ordered and understood as set of sequences, gives all possible paths through the rule set. One path is a sequence of rules $(r_1, r_2, ..., r_n)$. Running such a rule sequence means successive evaluation of each rule and, hence, a stepwise system-state modification:

$$I(M, \{r_1, r_2, ..., r_n\}) = I(I(M, r_1), \{r_2, ..., r_n\}) \text{ with } I(M, \{\}) = M$$

Conceptually, our reasoning engine constantly loops through a list of all goals, sorted by weight, and searches for a sequence of rules that transforms the current system state into the goal state, as specified by the guard of the goal. Whenever the system has to decide between two different rules that add the requested proposition to the memory, it uses a local-best strategy and tries the rule with the higher weight first.

13.2.2 Weight Model

In the preceding subsection we presented the semantic model of the runtime excluding details of weight calculation. We now address this shortcoming. Weight calculation naturally is based on a quantitative model taking past experiences, activity, and user's preferences into account. Thereby weight is calculated by some function $\gamma(\cdots)$ and defined over the interval [0...1] with a saturating behavior. The following parameters are needed by γ for weight calculation: (A) The current activity (*activity*) of a rule, (B) experiences learned from past runs of the rule ($damping$), (C) some max-activity point (max) and (D) some user-supplied function $\alpha(activity)$ that takes the current activity of the rule as input parameter. As already mentioned in the previous section, we define:

$$\gamma(activity, damping, max, \alpha) = \\ \alpha'(activity) \cdot Abs(max - activity) \cdot (1 - damping)$$

where $\alpha'(activity)$ is the first derivative of some function $\alpha(activity)$. α must not have arbitrary slope as $\alpha'(activity) \cdot (max - activity)$ has to be smaller than or equal to one in order for γ to stay within bounds. $Abs(max - activity)$ returns the distance to the user-defined maximum activity in a linear way. Finally, the last part of γ deals with experiences gained from past runs of the rule. It uses a damping factor ($damping$) which is a counter of failed attempts to execute the rule after the guard of the rule was determined to be true. As already indicated, $damping$ is defined in the range $[0...1]$ and we say $1 - damping$ to be the desirability.

Both, the damping- and the activity factor of a rule are determined by the system during runtime in a predefined way. Function α and the maximum activity max of a rule have to be provided by the rule designer and serve for the purpose of influencing rule selection in two different ways: The maximum value max is used to specify how often a rule should be selected over a given time frame whereas the slope of α is used to change, i.e. boost, the weight over a specified (user-selected) activity range. In this way α helps deciding between different rules reaching the same goal in case max of both rules has the same value. (α allows the developer to change the slope of $max - activity$ over a certain range.) We do not especially demand for α to have its maximum at the place of max but we demand a slope $>= 1$ until the specified max. As already mentioned, the additional requirement is that $\alpha'(activity) \cdot (max - activity) <= 1$ in order for $\gamma()$ to stay within bounds. As we also stated that γ is saturated, place-wise breach of this requirement won't harm.

After we have shown how to calculate the weight of a rule, we give the rationale behind. We assume that all rules stored within the rule set help the system advance to a good state. In other words, if the guard of a rule is satisfied it is beneficial for the system to choose the rule and run it. Whenever the guard of a rule is satisfied, we assume that all actions attached to the rule will execute without error. Of course the system monitors action execution and will increase the damping factor if it encounters an unexpected error while executing the selected rule. (No error means decreasing the damping factor if it is greater than zero.) This way transient- and permanent faults are masked from occurring too often. In our opinion, the developer will have different levels of activity in mind for different rules. In our system we use user supplied max values to implement this behavior: Depending where the user puts the maximum, the system will run the associated rule with a given frequency. We want to emphasize that no rule can be totally blocked from execution in order to prevent the system to run out of run-able rules.

This methodology was inspired by biological systems, like cells, that decode DNA sequences to build up useful proteins at different rates all the time when some part of the DNA is 'activated'. In cells there exist multiple ways of influencing the rate in which the DNA gets transcribed and because of this flexibility cells can react in very complex ways to external influences. In our rule-based system, we also want to have the flexibility to change frequency patterns at runtime. The most natural way of providing this functionality is by allowing rules to interact with the weight calculation function γ in defined ways. As a first step, we allow special rules to change the weight of other rules. This can be extended to changing user supplied parameters, like max and α. Rules that have this capability are called "*Meta-Rules*".

After presenting the basic idea behind the system's design, we give further mathematical properties of our system in the following section.

13.3 Additional Properties of the Weight Model

We have shown that the total weight of a rule depends on four factors: (A) Activity Factor, (B) Damping Factor, (C) α (*activity*), and (D) a maximum of activity (normally taken from α). Because most of these factors are somehow dependent on the activity, we start describing further mathematical properties of the system by discussing the activity factor. After presenting details on activity factor calculation, we continue presenting mathematical properties of the system by discussing the damping factor. We will answer the question how different damping factors relate to each other in terms of cross-over point. We do not include the proofs of the properties in this chapter. Interested readers find these in our technical report [2].

13.3.1 Properties of the Activity Factor

As already said, the system increments the activity factor when a rule has been chosen. It is, however, obvious that the activity factor cannot be increased endlessly: We need an aging operation that guarantees that the activity factor stays within bounds. Because division by two is a shift in binary digit based systems, we chose to age the activity factor periodically by dividing it by two. So the activity of a rule at iteration number *n* can be calculated as follows:

$$activity_n = \frac{1}{2} \cdot chosen_n + \frac{activity_{n-1}}{2}$$

where $chosen$ is either 1 or 0 depending on whether the rule was chosen or not and n is the iteration number. As we would like to run the system forever, $n \to \infty$. From the series representation of the formula above, it follows that the activity factor can at most be one. This is the case when the system runs forever and the rule is always selected.

Because the main idea behind the presented approach is to specify the frequency of how often a given rule or goal should be run over a particular time frame, i.e., the running time of the system, we have to answer the question of where to put the maximum of $\alpha(activity)$. Putting the maximum at the right position will guarantee that the rule is selected with the chosen frequency. In the following, the natural number *frq* defines the length of an interval the total number of iterations is divided by. In other words, *frq* says that the rule is selected every *frq*th iteration. For example, when set to one, the rule is always selected. If *frq* is set to two, the rule is selected only every second time.

Let $frq \in N$ and $frq > 0$ be the length of the interval defining how often a rule is chosen. If the system runs forever ($n \to \infty$), the maximum value of the activity factor can be calculated as follows.

$$activity_{Max}(frq) = \frac{2^{frq}}{2 \cdot (2^{frq} - 1)}$$

Given this result, we want to know how many steps n are needed in order to be close to the limit. It can be shown that in order to get $1/\varepsilon$ ($\varepsilon > 0$) close to the maximum, one needs *n* steps, with *n* given by (*frq*>0)

$$n = ld(\frac{\varepsilon}{2^{frq} - 1})$$

Because the activity value gets a boost when the rule is selected and is aged afterward, there exists a minimum value of the activity factor too. The minimal activity value for a given selection pattern frq can be calculated as follows

$$activity_{Min}(frq) = \frac{1}{2^{frq} - 1}$$

The difference between maximum and minimum activity is given by the bandwidth:

$$activity_{\delta}(frq) = \frac{1}{2} - \frac{1}{2 \cdot (2^{frq} - 1)}$$

The bandwidth is dependent on frq and becomes 1/2 if $frq \to \infty$ and 0 when $frq = 1$. Consequently, if one knows the maximum and minimum activity values, one can use this knowledge to calculate the desired value of *frq*:

$$frq = ld(\frac{activity_{Max}(frq)}{activity_{Min}(frq)} \cdot 2)$$

We know that the lower limit of $activity_{Max}$ ($frq \to \infty$) is 1/2 and the upper limit is equal to one. We also know that a small value of *frq* results in an activity maximum near one, whereas for big values of *frq* the activity maximum settles near 1/2. Given this lower limit, we are interested in the maximum value of *frq* we can still represent with a given number of decimal places: With a given precision of $1/\varepsilon$ we can represent all activity maximum values of a given number *frq* up to

$$frq \leq ld(\frac{\varepsilon + 2}{2}) \text{ and } e + 2 > 0$$

This means that with a precision of 5 digits ($\varepsilon = 100000$), we find that we can use all *frq*-values below 15. After discussing the activity factor, we now turn our attention to the damping factor.

13.3.2 Properties of the Damping Factor

Given two functions $\alpha_1(x), \alpha_2(x)$ with maxima $M_1, M_2 = M_1 + z$ ($z \in R+$) that are mutually exclusive, i.e., $\alpha_{a1}(x), \alpha_{a2}(x)$ represent two alternate rules achieving the same goal, we want to know at which point $q = M_1 - \Delta$ the weights are equal, in other words, $\gamma_1(q) = \gamma_2(q)$. This is the point at which the system will switch from using one rule to the other one. The weight of rule one is calculated as follows:

$$\gamma_1(x) = \alpha_{a1}'(x) \cdot (M_1 - x) \cdot d_1$$

with $x = activity_{n,\gamma_1}$ and $d_1 = 1 - damping_{\gamma_1}$

Weight calculation for the second rule is done equally. As we want to find the equilibrium of the weights, we set equal

$$\alpha_{a1}'(x) \cdot (M_1 - x) \cdot d_1 = \alpha_{a2}'(x) \cdot (M_1 + z - x) \cdot d_2$$

With $x = M_1 - \Delta$ and in the special case of $\alpha_{a1}' = \alpha_{a2}'$ we derive

$$\frac{d_1}{d_2} = \frac{z + \Delta}{\Delta}$$

Most of the time, however, we are interested in the number of failed attempts to run one rule before the system will switch to the alternative one. If we assume $1 = M_i \cdot S_i$ then $1 - damping = d = 1 - f \cdot S_i = (M_i - f) \cdot S_i$ where f represents the number of times a rule failed and S_i is an increment factor. From

$$\frac{d_1}{d_2} = \frac{(M_i - f_1) \cdot S_i}{(M_i - f_2) \cdot S_i} \text{ it follows that } f_1 = M_i - \frac{d_1}{d_2} \cdot (M_i - f_2)$$

Of course the damping factor relates to the probability that all actions within a rule can be carried out: If we use P_f for the probability of the actions to fail and $P_o = 1 - P_f$ as probability for action completion, we can show that $damping = f \cdot S_i \cdot (P_f - P_o)$. If we assume a stationary fault we can distinguish between two different modes of behavior: If $(P_f - P_o) > 0$ then the damping factor will increase over time ($n \to \infty$) until its saturation value 1. In principle this means that if a rule's actions fail more often than they succeed, the rule is (mostly) blended out over time. Otherwise $(P_f - P_o \leq 0)$, the damping factor will – in sum – be decremented to zero.

13.4 Results

We have implemented the presented approach on a hardware prototype of our remote sensing platform. The device uses a Microchip PIC 18F controller that provides 128 kBytes program memory and about 4 kBytes random access memory. Because of these restrictions and the need to include firmware for the driver layer, our implementation works with a fixed α function: Only the *max* values of the rules can be set by the developer.

Table 13.1 Permanent Faults. "G1" sends acquired position data by GSM, while "G2" sends the information by some near-field communications link. An attached "*x*" means the goal was selected but failed. Besides using GPS, the system can also acquire position information by using information provided by a compass module. "GP" is the power-save goal

	Scenario A		Scenario B		Scenario C	
Iteration	Goals	Sim. Faults	Goals	Sim. Faults	Goals	Sim. Faults
1	G0, G1	–	G0, G1	–	G0, G1	–
2	G2, G0	–	G2, G0	–	xG2, G0	GPS
3	G1, G0	–	G1, G0	–	xG1, G0	GPS
4	G2, G0	GSM	xG2, G0	GPS	G0, xG2	GPS,GSM
5	xG1, G0	GSM	xG1, G0	GPS	G0, xG1	GPS,GSM
6	G2, G0	GSM	G0, xG2	GPS	G0, xG1	GPS,GSM
7	G2, G0	GSM	G0, xG1	GPS	G0, G2	GPS,GSM
8	G2, G0	GSM	G0, xG1	GPS	G0, G2	GPS,GSM
9	G2, G0	GSM	G0, G2	GPS	G0, G2	GPS,GSM
10	G2, G0	GSM	G0, G2	GPS	GP, G0, G2	GPS,GSM, LowBatt

Tables 13.1 and 13.2 present a case study and illustrate the capability of the system to adapt to changing environments. The rule set used during this experiment is quite simple and includes two goals (G1, G2) for sending position information. Note that these two goals are mutually exclusive, meaning the system will either select G1 or G2. Further goals are included for switching into power save mode (GP), leaving power save mode, and for shutting down the system in case of low energy conditions. The results of Table 13.1 are interesting to compare because they demonstrate the system is quicker in adapting to a single fault than to multiple ones that tend to cause fuzzy responses. Please note that our example implementation had a policy of sticking to GPS as long as possible and to ignore a certain amount of failures of this module. Therefore the system needs more time to adapt to the non functional GPS module than it needs when adapting to a faulty GSM module. Table 13.2 shows a more complex set of induced faults. Note again that the system has a policy of using GPS for a certain amount of time, even if it seems faulty. In general the experiment shows three parts. The first part includes iterations one to five. This part simulates transient, changing faults with multiple influences on goals.

Table 13.2 Transient Faults. "G1" sends acquired position data by GSM, while "G2" sends the information by some near-field communications link. An attached "*x*" means the system fails to reach the goal. Also see Table 13.1

	Scenario A		Scenario B	
Iteration	Goals	Sim. Faults	Goals	Sim. Faults
1	G0, xG1	GSM	G0, xG1	GSM
2	xG2, G0	NF	xG2, G0	NF
3	G0, xG2	NF, GPS	G0, xG2	NF, GPS
4	G0, G1	NF	G0, G1	NF
5	G0, G1		G0, G1	
6	G0, G1		G0, xG1	GPS, GSM
7	G0, xG1	GPS	G0, xG2	GPS, GSM
8	G0, xG2	GPS	G0, xG1	GPS
9	G0, xG1	GPS, GSM	G0, xG1	GPS
10	G0, xG1 (pos w. GPS)	GSM	G0, xG2	GPS, GSM
11	G0, G2		G0, xG1 (pos wo. GPS)	GSM
12	G0, G2		G0, G2	
13	G0, xG2	NF	G0, xG2 (pos. wo. GPS)	NF
14	G0, G1	NF	G0, G1	NF
15	G0, G2		G0, xG2 (pos wo. GPS)	NF
16	G0, G1		G0, G1	NF, GPS
17	G0, G2	Compass	G0, xG1 (pos wo. GPS)	NF, Compass, GSM
18	G0, G1		G0, G1 (pos w. GPS)	NF, Compass
19	G0, G2		G0, xG2	NF
20	G0, G1		G0, G1	NF

The second part of the experiment is a sequence of GPS faults. As we know that the system has the policy of strongly preferring GPS over its backup it comes as no surprise that the system can't reach a few goals in sequence. As the system does not infer *why* it was unable to reach the goals, it keeps switching alternatives. This part of the experiment therefore recommends including techniques to reason about a fault. However, our prototype system (and the main field of application) is geared toward small devices and low energy consumption which is in contrast to our wish of expanding the presented approach significantly. The last part of the experiment finally shows how our system adapts to a persistent fault of the NF component. Note that the experiment was carried out in a way that the activity factor is only increased when the goal is reached. If the activity factor is increased whenever the system tries to reach a goal, the device is more successful and chooses seven out of nine cases correctly, as further experiments have shown. Before concluding, we want to add that the frequency specified by α has been set to "always" for all goals. Therefore the system tries to reach either G1 or G2 in all iterations. G0 is a goal for house-keeping functionality: It assures that the system is in on-state.

13.5 Conclusion

The presented methodology provides ways to specify preferred and backup behavior, and uses damping factors to discriminate between permanent and transient faults. The approach has the ability to automatically switch back to the preferred behavior, and can be implemented on resource constrained devices. In this chapter, we discussed the principal architecture and the semantic of our rule-based approach.

Acknowledgements This work has been supported by the FIT-IT research project SEPIAS. The first author is currently funded by MOGENTES, a Specific Targeted Research Project (STREP) in the 7th EU Framework Programme.

References

1. R. A. Brooks. Intelligence Without Reason. In J. Myopoulos and R. Reiter, editors, Proceedings of the 12th International Joint Conference on Artificial Intelligence (IJCAI-91), pages 569--595, Sydney, Australia, 1991. Morgan Kaufmann publishers Inc.: San Mateo, CA, USA
2. W. Krenn, F. Wotawa. The SEPIAS Embedded Runtime – Single System Semantics. Technical Report SEPIAS-TR-2008-01, Graz University, 2008.
3. N. J. Nilsson. Teleo-Reactive Programs for Agent Control. Journal of Artificial Intelligence Research, 1:139–158, 1994. AAAI Press
4. J. K. Rosenblatt and D. W. Payton. A fine-grained alternative to the subsumption architecture for mobile robot control. Proceedings of the IEEE International Conference on Neural Networks, pages 317–324, 1989. IEEE Press
5. A. Saffiotti, K. Konolige, and E. H. Ruspini. A multivalued-logic approach to integrating planning and control. Artificial Intelligence, 76(1–2):481–526, 1995.
6. W. Krenn and F. Wotawa. Configuring collaboration of software modules at runtime. In Configuration – Papers from the 2007 AAAI Workshop, pages 19–24, Vancouver, Canada, 2007. AAAI Press

Chapter 14

Digital Open-Loop Control of a Piezoelectric Valve for Household Appliances

Daniele Petraccini[1], Massimo Conti[1], Fortunato Nocera[2], Lorenzo Morbidelli[2] and Fabrizio Concettoni[2]
[1]*D.I.B.E.T., Università Politecnica delle Marche, Ancona, Italy;*
[2]*Indesit Company S.p.A., Fabriano, Italy, m.conti@univpm.it*

Abstract Piezoelectric materials are widely used as sensors and actuators in many applications. They allow efficient digital control of mechanical systems, but suffer from nonlinearity and hysteresis. This chapter presents a new digital open-loop control of piezoelectric bender for real-time applications in domestic appliances, for which strong specifications are good accuracy and low cost. The control algorithm, implemented in a microcontroller, solves the problems of nonlinearity and hysteresis. The chapter presents experimental results of the prototype that has been realized in the Indesit Company laboratories for future applications on domestic appliances.

Keywords Piezoelectric actuator; Hysteresis; Open-loop control

14.1 Introduction

Since the discovery of piezoelectric effect in 1880 by Pierre and Jacques Curie, scientific knowledge and technology of piezoelectric materials have dramatically improved so that nowadays they are used in a huge number of applications, working as sensors, transducers, and actuators.

In particular, piezoelectric devices are applied in embedded systems in several different areas: automotive (fuel injectors), consumer electronics (auto-focus in reflex cameras), computer accessories (ink-jet printers), music (pickups for electrically amplified guitars, electronic drum pads, and loudspeakers), medical (ultrasonic transducers for medical imaging), optics, and so on.

N. Martínez Madrid, R.E.D. Seepold (eds.), *Intelligent Technical Systems*, Lecture Notes in Electrical Engineering 38, DOI 978-1-4020-9823-9_14,

In the field of domestic appliances, piezoelectric materials are used as well, mainly as sensors. For example a touch interface can be realized that is based on direct piezoelectricity: when the finger applies a slight pressure on the sensor, a voltage is generated. Such technology is integrated into the user interface of a model of Ariston electric oven that is currently on the market (see also Ariston web-site [1]).

This chapter investigates the possibility of using piezoelectric actuators in domestic appliances as an alternative to classical actuators. Most of actuators currently used in household appliances are mechanically related to the action performed by the user. Examples are: thermostat of cooling appliances, knob of cooking appliances. An electro-mechanical piezoelectric device would allow placing the real actuator in a location which differs from the interface on the outside, where the user performs a command.

Our work focuses on a bender piezoelectric actuator, which is a cantilever beam made up by two layers (one ceramic layer and one supporting layer) that bend when a voltage is applied: this happens because under the action of the electric field, piezoelectric layer shrinks and forces a deflection of the whole bender. The piezoelectric bender that we used allows deflection in one direction only; two directions bender can be realized by adding a second piezoelectric layer on the other side of the supporting layer.

Notice that the bending element can be used in a variety of ways according to the mechanical system that is connected to the cantilever.

We chose to experiment with open-loop control of piezo bender, in order to reduce complexity and cost, so as to make it more feasible to be applied. Moreover, in some cases a closed-loop control cannot be implemented at all since the required sensor should prove too costly or difficult to be put into action.

Unfortunately, piezoelectric materials suffer from hysteresis, a form of nonlinearity containing memory of history. This major limitation makes them more difficult to be controlled and requires some form of compensation to be implemented.

Some open-loop techniques based on models proposed in the literature were tested with the actual piezo bender, but the results achieved did not completely satisfy authors. For these reasons, we developed a novel empirical technique which proved to be effective in the case of stepwise variables.

The general problem of hysteresis in piezo actuators is discussed in Section 14.2. Then, Section 14.3 describes the main components of our prototype. Two open-loop techniques are outlined in Section 14.4, and in Section 14.5 experimental results are provided, followed by conclusions.

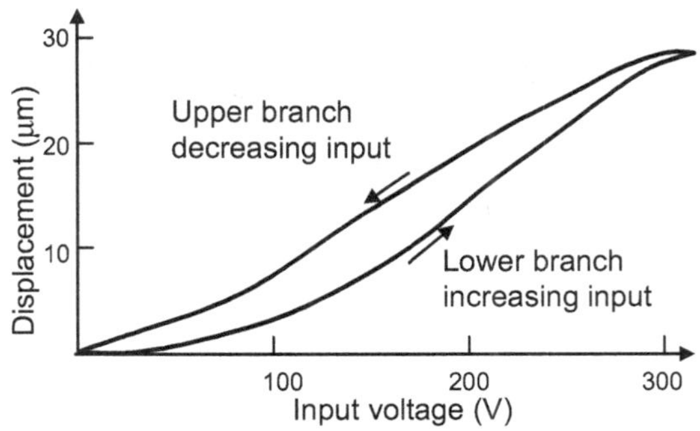

Fig. 14.1 Typical displacement-voltage characteristic of a piezoelectric actuator

14.2 Hysteresis in Piezoelectric Actuators

A piezoelectric actuator is an electrically controllable device which works on the basis of the piezoelectric effect: when subjected to an externally applied voltage (input), the material changes shape by a small amount and this deformation can be used in a variety of ways. The main advantages of piezoelectric actuators are their high stiffness, fast frequency response, and high resolution – properties that give reason for their vast application.

Unfortunately, they suffer from hysteresis: the entity of the deformation (output) of a piezo actuator does not depend uniquely upon the present value of the input voltage, but is influenced by the history of input, in particular by the sequence of past voltage extrema.

Figure 14.1 shows the typical hysteretic characteristic of a piezoelectric actuator. The curve is obtained by driving the actuator with a voltage starting from 0 (null exitation) and continuously increasing up to 300 V: the corresponding displacement follows the lower branch of the hysteresis loop. When the input voltage is decreased from 300V down to 0, the displacement follows the upper branch of the hysteresis loop. This behaviour is obviously undesirable since the displacement resulting from a voltage input signal V(t), showing a series of local minima or maxima, becomes in general unpredictable.

Several approaches have been proposed in the literature to compensate for such an unwanted effect, and they can be classified into the following categories [2].

1. Electric charge control. As Newcomb and Flinn suggested [3], the linearity of piezoceramic actuators can be significantly improved if an electric charge – instead of a voltage – is applied and varied to control the deformation. Nevertheless, electric charge control needs specially

designed charge amplifier, and good linearity cannot be guaranteed in a wide frequency range.

2. Closed-loop displacement control. Typically, strain gauges are used as feedback sensors. Examples of this kind of approaches can be found in [4], where a linear control technique is used after linearization of the hysteresis; in [5], where adaptive control with an approximate model of hysteresis is used; in [6], where a neural network is used to learn the nonlinearity. Anyway, such a strategy requires an additional cost on the device and this may be undesirable since the cost of each piece of the machine is crucial in the world of consumer electronic appliances.
3. Linear control with feed forward inverse hysteresis model (open-loop control). In this approach the idea is to find a proper model of the hysteretic behaviour of the piezo actuator and use its mathematical inverse in the control chain. Examples of such approach can be found in [7,8], where the inverse Preisach model is applied for hysteresis cancellation; in [9] where the realization of an inverse feedforward controller for the linearization of a piezoelectric device is formulated.

We tried to perform an open-loop control of the piezoelectric bender, which required the hysteretic behaviour of the device to be well-characterized.

14.3 Prototype Description

A significative example of a piezoelectric actuator, suitable for domestic appliance can be presented by valves for fluids control. Figure 14.2 provides a schematic illustration of the prototype of piezoelectric valve used in laboratory.

The valve, devised for controlling a fluid, comprises a hollow body having a fluid inlet duct and a fluid outlet duct. The cavity of body is divided by a rigid separator into an upper outlet cavity and a lower inlet cavity. Separator has an aperture (a flared hole) allowing the fluid to flow from lower cavity to upper cavity. Shutter is adapted to shut aperture of separator by perfectly coupling thereto, thus ensuring tightness. Body of the valve has a hole in its upper portion which allows the stem to slide without any substantial fluid leakage from chamber.

The piezoelectric bender acts on the regulating device, composed by the shutter and the stem, rigidly joined together. The actuator receives an electric signal through two electric leads; actuator is so conceived that, when a direct electric voltage is applied, it will curve downwards and the regulating device will move downwards accordingly; the greater the amplitude of the voltage signal, the more actuator will curve.

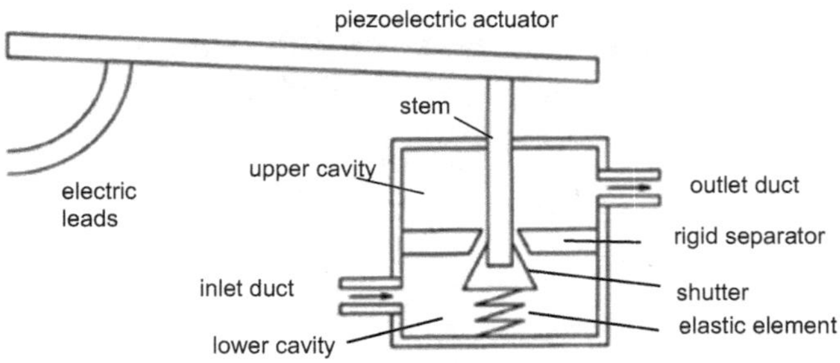

Fig. 14.2 Schematic representation of the piezoelectric valve

When the input signal has a null voltage value (or a value below the minimum design voltage value), the valve is "fully closed" because shutter is shutting aperture; this is the idle or inoperative condition of the valve. When the input signal has the maximum design voltage value, the valve is considered to be "fully open" because shutter is far below aperture and therefore cannot prevent the fluid from flowing from lower chamber to upper chamber. An elastic element, in particular a spring, is provided in order to obtain or facilitate the return of the valve to its idle or inoperative condition.

The test prototype, realized in Indesit Company laboratories, consists of the piezoelectric valve, the electronic control system, the meter used to measure the flow and an interface with a PC used to program the electronic control system and to monitor the results. The desired regulations of the valve are provided by the user through a software program running on a PC. The generated input pattern is transmitted to the electronic board, which includes all the circuitry needed to elaborate the input signal and to supply the piezoelectric actuator with the proper voltage. From the circuit point of view, the piezo actuator can be thought of as a capacitor, whose applied voltage determines the entity of the mechanical deformation.

The piezo actuator carries out a mechanical action on the valve. The meter block represents a gauge that provides the measure of the flow to the acquisition system.

The electronic board, in Fig. 14.3, comprises three main functional blocks:

- the power supply module, consisting of a switch mode power supply, generating the 5V dc power supply for microcontroller and the 300V dc power supply needed to synthesize the piezo control voltage.
- the control module, consisting of a microcontroller which implements the following: the hysteresis compensation technique described later in details; the communication with the user interface based on a standard RS-232 serial line with an Indesit proprietary communication protocol; the decoding of the input pattern and the conversion into the corresponding voltage pattern to be generated; the generation of the control signal for the piezo actuator interface.

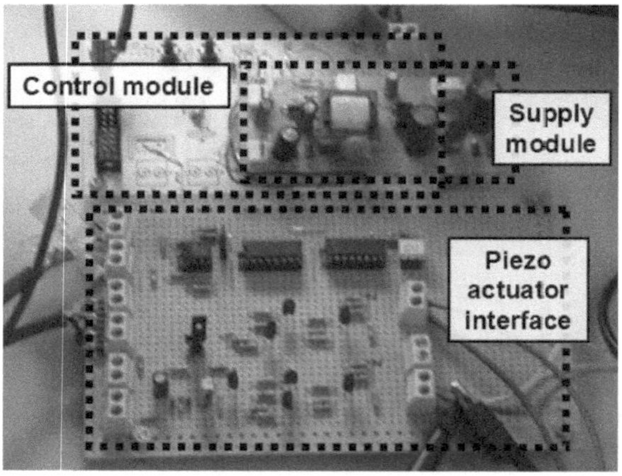

Fig. 14.3 The electronic board and its main modules

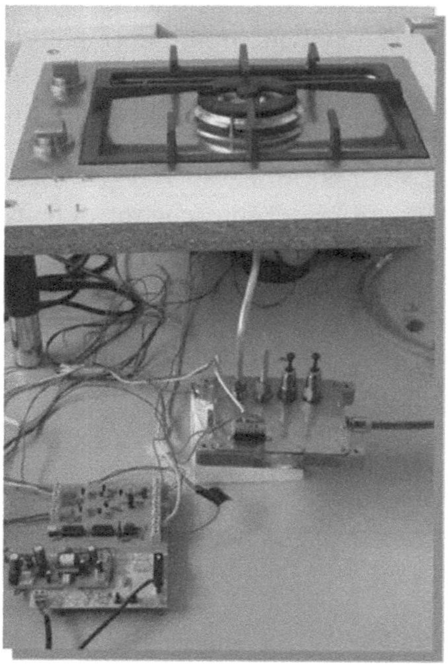

Fig. 14.4 The electronic board applied to the control of a test gas cooking hob

Fig. 14.5 Future application of piezoelectric valve to gas cooking hobs

Such control signal is a Pulse Width Modulated signal whose duty-cycle determines the piezo driving voltage. In order to generate the required voltage, a closed-loop control system is used: the microcontroller measures the actual voltage applied to the piezo and varies the duty cycle of the PWM signal according to a PID control algorithm. The PID has been tuned in order to satisfy the required response time and noise rejection. The PID approach was chosen in order to have a versatile tool to adapt the piezo behaviour to different application requirements.

- the piezo actuator interface is a circuitry that adapts the 0–5V PWM signal from the microcontroller to the 0–285 V range and, by means of a low-pass filtering, generates a continuous voltage in the range up to 285 V dc. The voltage applied to the piezo is provided, with an appropriate scaling factor, to the microcontroller as a feedback signal.

The electric control board has been applied to the control of a gas cooking hob, as shown in Fig. 14.4, for future application on commercial gas cooking hob of the type shown in Fig. 14.5.
The output signal of our system, the gas flow, is acquired while the input voltage is gradually increased from 0 V up to 285 V and then gradually decreased down to 0 V. The hysteretic behaviour of the piezo is reflected to the characteristic of the whole cooking hob.

In a typical operating condition of the valve, the input voltage changes according to the input pattern provided by the user. As a result of the memory inherent in the piezoelectric device, the value of the output at a given time is difficult to be predicted, since it depends on the sequence of past input values.

In general, the operating point in the input-output diagram at a given time lies inside the major hysteresis loop, but its exact location is unknown if we assume that the output is not observable. Hence the area enclosed in the major hysteresis loop gives a measure of the degree of uncertainty associated to the system.

Obviously, this highly nonlinear behaviour makes the actuator difficult to be driven in an open-loop scheme, and requires a proper compensation technique to be applied.

14.4 Hysteresis Compensation Technique

With reference to the schemes for hysteresis compensation described in Section 14.2, we decided to use a feed forward inverse hysteresis model (open-loop control). In particular, from the literature we borrowed the classical Preisach model [7,8] and the more recent model of Prandtle-Ishlinskii [10,2].

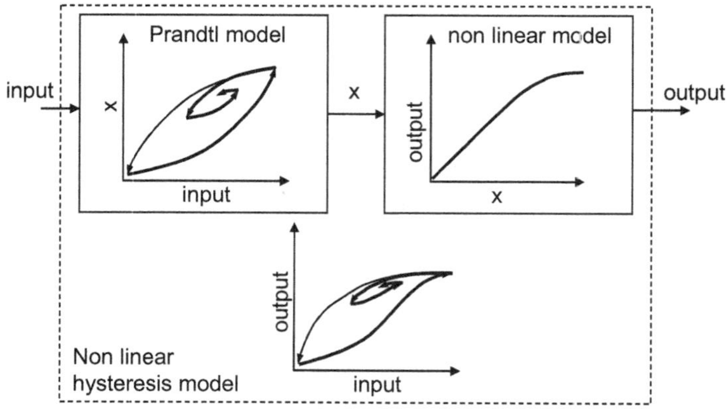

Fig. 14.6 Non linear hysteresis model

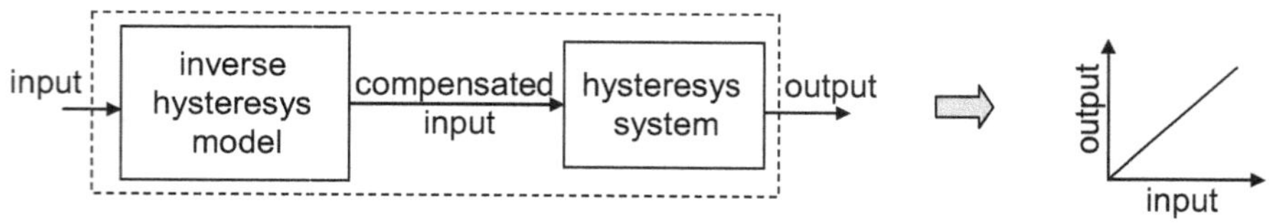

Fig. 14.7 Hysteresis cancellation with inverse model

Preisach model cannot be applied in our case since the wiping-out and congruency properties are not verified, hence necessary and sufficient conditions for the representation of the actual hysteresis nonlinearity are not satisfied. As regards the model based on Prandtle-Ishlinskii operator, we found out that it doesn't work as well: indeed, a correct identification of the hysteresis curve cannot be accomplished due to the strong nonlinearity of the characteristic and the concavity of both the upper and lower branches of the main loop.

Hysteresis compensation with a non linear hysteresis model

We modified the Prandtle model introducing a piecewise nonlinearity at the output of the hysteresis model, as shown in Fig. 14.6. In this way complex shapes of the hysteresis systems can be modelled. This nonlinear hysteresis model is, in our knowledge, original and never proposed by other researchers. We defined and algorithm to estimate the coefficients of both models by fitting the experimental data. The non linear hysteresis model can be used to identify the inverse model. The compensated input signal applied to the hysteresis system for hysteresis cancellation will generate the desired linear input-output relationship, as shown in Fig. 14.7.

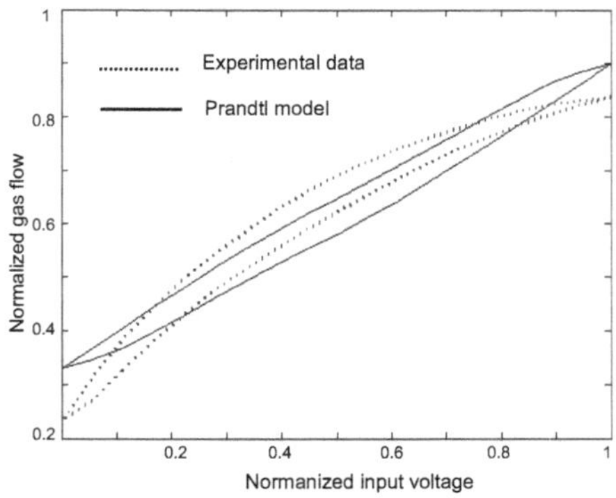

Fig. 14.8 Experimental hysteresis and Prandel best fitting model

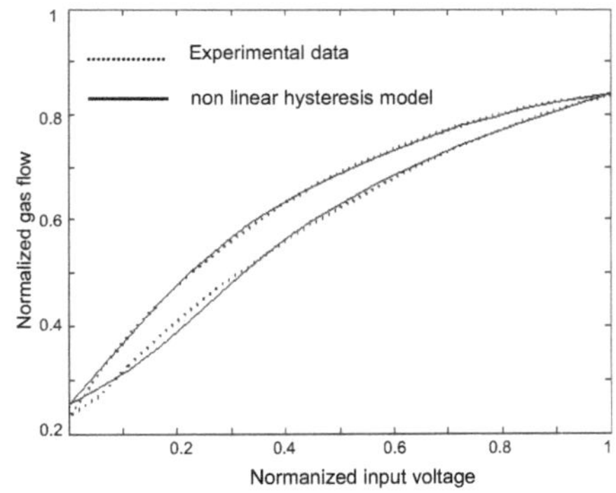

Fig. 14.9 Experimental hysteresis and nonlinear hysteresis best fitting model

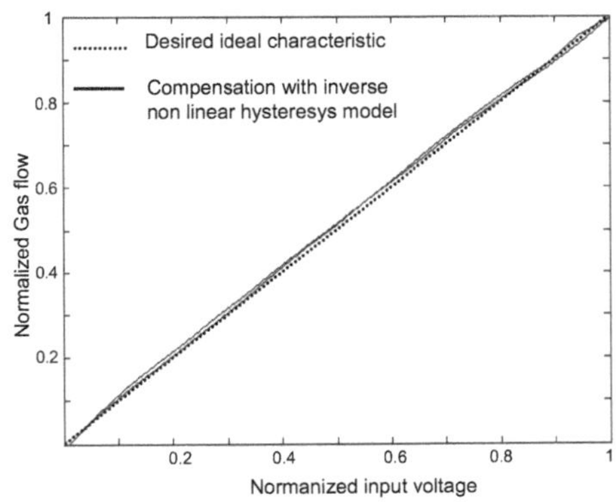

Fig. 14.10 Hysteresis cancellation with inverse nonlinear hysteresis model

Figure 14.8 shows the experimental data and the best fitting obtained by the Prandtle model of order n=20, while Fig. 14.9 shows the best fitting obtained by the nonlinear hysteresis model with a Prandtle model of order n=20 and the piecewise linear curve of order m=15. The improvement is evident.

The hysteresis cancellation obtained with the inverse nonlinear hysteresis model with a Prandtle model of order n=20 and the piecewise linear curve of order m=15 is shown in Fig. 14.10.

Upper branch control technique

The nonlinear hysteresis compensation technique previously described requires the implementation of the inverse model of the nonlinear hysteresis in the microcontroller, the complexity of the model forces the designer to high performances and high cost microcontrollers. For industrial high volume applications, the statistical variations of the piezo characteristics must be carefully considered; a tuning of the inverse model could be necessary for each product. These considerations limit the real applicability of the nonlinear hysteresis compensation technique. To overcome these problems, we have developed an empirical technique which proved to be effective in the case of stepwise variables, that is when the desired time behaviour of the output is a series of levels or steps.

Our technique exploits the fact that the piezoelectric actuator has a fast frequency response and can be driven by the electronics very rapidly, being such electronics embedded in the whole appliance system. In particular, our developed circuitry can make the piezo to go from the minimum up to the maximum voltage level in about 50 ms. At the same time, the driven system (fluid dynamic, in this case) can be seen as a dynamical system with an upper cut frequency that is certainly lower than the frequency at which a piezo device can be made to operate.

We can take advantage from this fact by biasing the actuator to the upper branch of the hysteresis curve, hence making it to work in a well defined path. This can be achieved by applying a transformation to the stepwise input signal as follows:

- Every step-down transition ($V_H \rightarrow V_L$) of input signal is left unchanged;
- Every step-up transition of input signal ($V_L \rightarrow V_H$) is converted to a combined transition ($V_L \rightarrow V_{MAX} \rightarrow V_H$), where the value V_{MAX} is the maximum applicable voltage (e.g. 285V in our case) and it is kept for a time interval as small as possible (e.g. 50 ms).

In the latter condition, the apparatus driven by the piezo can't react to the rapid transitions $V_L \rightarrow V_{MAX} \rightarrow V_H$, due to its inertia, and behaves like if just the $V_L \rightarrow V_H$ transition is happening.

In such a manner, the stable operating points of the piezo actuator lie in the upper branch of the hysteresis curve, resulting in a nonlinear hysteresis-free behaviour. The remaining nonlinearity – which is memory-less – can easily be compensated, for example by means of a look-up table in the final control algorithm.

The foregoing technique has been implemented in a 64-pin microcontroller operating at 8 MHz; the complete application – voltage control and algorithm implementation – required more than four Kbytes of assembly code.

The details of the modifications needed to the input signals and the effectiveness of the resulting approach on the output signal are better illustrated in the following section, where an experimental case is described.

According to the explained principle, a compensation of hysteresis could be obtained by making the piezo to operate in the lower branch of the hysteresis curve: in this case every step-down transition would require an additional intermediate step down to a voltage that causes the piezo to turn back to the no-deflection state, that corresponds to a null output. Anyway it is preferable not to pass through the null output, since it could have a negative impact on the valve behaviour, as we experienced with an alternative version of our prototype.

14.5 Experimental Results

This section discusses a representative experiment conducted with our developed prototype, whose main parts are illustrated in Section 14.3.

Figure 14.11 depicts the pattern used as input in our test. Seven different levels were employed, and the input sequence was conceived so as to set a given level several times during the test and through different transitions. Every step lasted 5 min, and a succession of 24 steps was used, for a total duration of 2 hours.

This sequence of levels was elaborated by the microcontroller, giving rise to the voltage signal depicted in Fig. 14.12. Notice the 285V spikes generated at each step-up transition of input pattern, according to the compensation technique described in Section 14.4.

The voltage signal generated by the electronic board was applied to the piezoelectric actuator. The resulting bender deformation produced a variation in the flow. The normalized output measured by the flow meter is plotted versus time in Fig. 14.13 (thick line).

Observe that throughout the test, the measured output levels are in very good agreement with the desired levels (horizontal dashed lines in the diagram).

In order to provide an idea of the improvements achievable with our technique, the values of the raw output without compensation are shown as well for each step (small diamonds): if the compensation algorithm is not applied, the resulting output level is not univocally associated to the input level, due to the large nonlinearity introduced by hysteresis.

Based on the time history of input and output signals, scatter diagrams can be constructed for the uncompensated and compensated cases, as shown in Figs. 14.11 and 14.12, respectively.

By comparing the input-output relationships in Figs. 14.14 and 14.15, the advantages of the proposed technique are clear: the compensation algorithm dramatically reduces hysteresis nonlinearity. In particular, the maximum error reduces from 9 to 2.8 % and the average error from 5.3 to 1.3%.

Such improvement in the overall system performances makes the open-loop control of piezo actuators promising for future application on the field of domestic appliances.

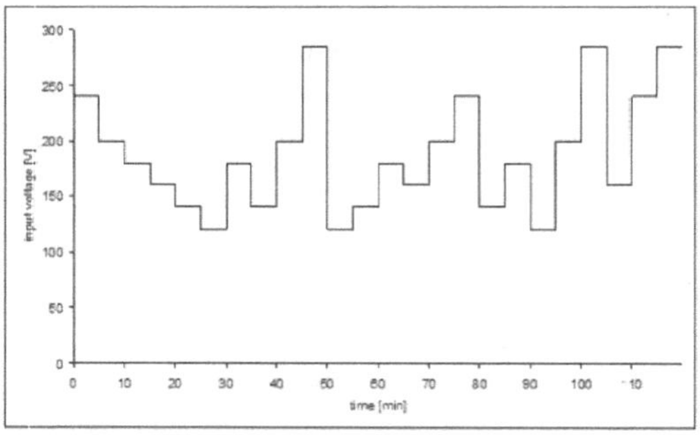

Fig. 14.11 Input pattern used in the test

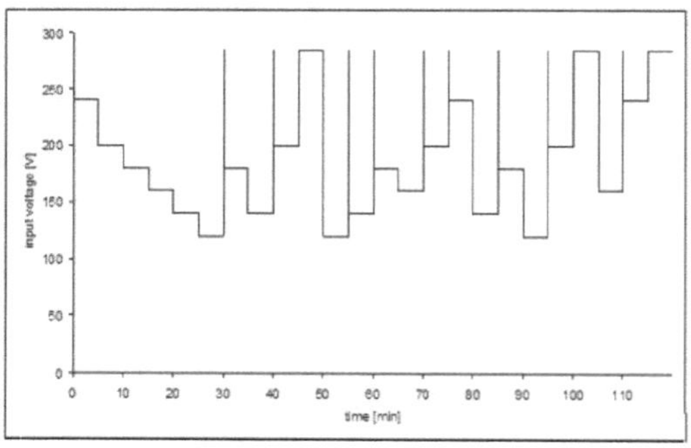

Fig. 14.12 Input signal provided by the controller

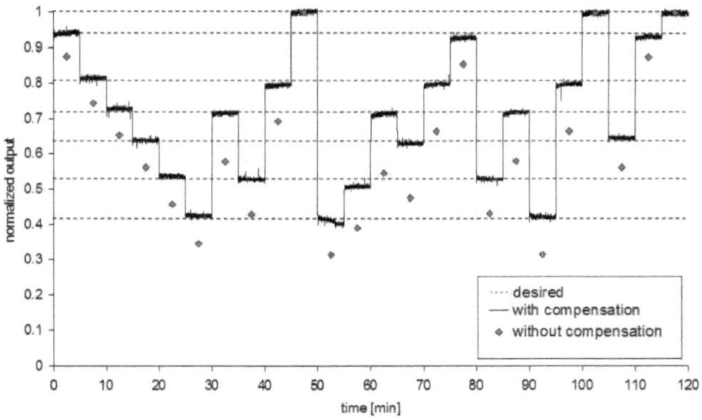

Fig. 14.13 Normalized output with compensation and without compensation

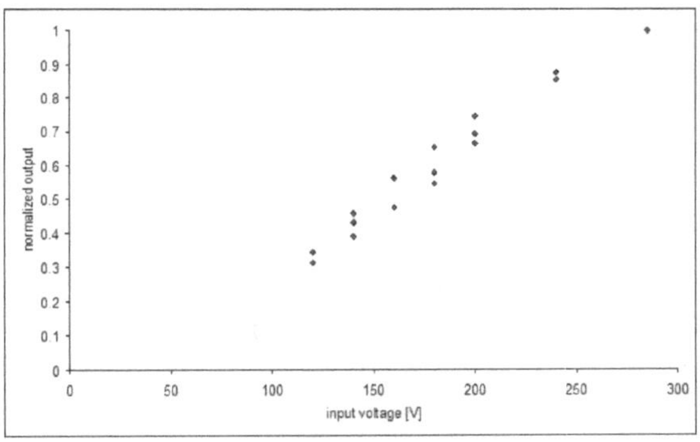

Fig. 14.14 Scatter diagram without compensation

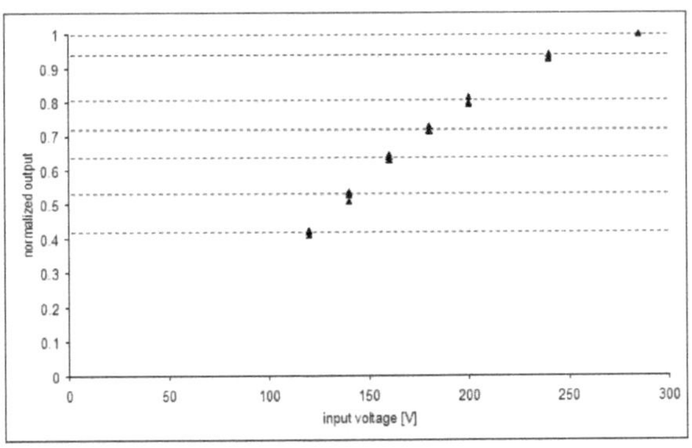

Fig. 14.15 Scatter diagram with compensation

14.6 Conclusions

This chapter presented a new digital open-loop control of piezoelectric bender for real-time applications in domestic appliances, where low cost is a fundamental requirement. Experimental results for the prototype realized in Indesit Company laboratories show the effectiveness of the proposed technique, which allows for a good hysteresis compensation; a sufficient precision seems to be gained in the investigated application, hence allowing the removal of direct feedback on the physical quantity to be regulated.

The illustrated technique is suitable to be directly implemented in the same microcontroller already present on currently marketed appliances.

The presence of piezoelectric actuators aimed at replacing traditional mechanical actuators, together with a more pervasive presence of embedded intelligence, enables the implementation of new functionalities, making it interesting for future applications on domestic appliances.

References

1. http://www.ariston.it/ariston/productsheet.do?productId=39354IT
2. Wei Tech Ang, F.A. Garmon, P.K. Khosla, C.N. Riviere. Modeling rate-dependent hysteresis in piezoelectric actuators. In Proceedings of the 2003 IEEE/RSJ, International Conference on Intelligent Robots and Systems, Vol. 2, 1975–1980, October 2003.
3. C.V. Newcomb and I. Flinn. Improving the linearity of piezcelectric ceramic actuators. Electronics Letters, Vol. 18, No. 11, pages 442–444, May 1982.
4. C. Jan and C-L. Hwang, Robust Control Design for a Piezoelectric Actuator System with Dominant Hysteresis. In Proc. 26th Conf. of the IEEE IECON2000, Vol.3, pp. 1515–20.
5. G. Tao and P. V. Kokotovic. Adaptive Control of Plants with Unknown Hystereses. IEEE Tran. Auto. & Control, Vol. 40, No. 2, pages 200–212, February 1995.
6. S.-S. Ku, U. Pinsopon, S. Cetinkunt, and S. Nakjima. Design, Fabrication, and Real-time Neural Network of a Three-Degrees-of-Freedon Nanopositioner. IEEE/ASME Trans. Mechatronics, Vol. 5, No. 3, pages 273–280, September 2000.
7. Y. Lv and Y. Wei. Study on open-loop precision positioning control of a micropositioning platform using a piezoelectric actuator. Fifth World Congress on Intelligent Control and Automation, 2004, Vol. 2, pages 1255–1259, June 2004.
8. G. Song, J. Zhao, X. Zhou and J.A. De Abreu-Garcia. Tracking control of a piezoceramic actuator with hysteresis compensation using inverse Preisach model. IEEE/ASME Transactions on Mechatronics, Vol. 10, Issue 2, pages 198–209, April 2005.
9. C.-H. Ru, L. Sun, and M.-X. Kong. Adaptive inverse control for piezoelectric actuator based on hysteresis model. In Proceedings of 2005 Internation*al Conference on Machine Learning and Cybernetics*, Vol. 5, pages 3189–3193, August 2005.
10. R. Changhai, S. Lining, R. Weibin, and C. Liguo. Adaptive inverse control for piezoelectric actuator with dominant hysteresis. In *Proceedings of the 2004 IEEE Int. Conf. on Control Applications*, Vol. 2, pages 973–976, September 2004.

Chapter 15

Comining Quantitative and Qualitative Models with Active Observtions to Improve Diagnosis of Complex Systems

Gerald Steinbauer and Franz Wotawa

Institute for Software Technology, Graz University of Technology, Inffeldgasse 16b/2, A-8010 Graz, Austria, steinbauer@ist.tugraz.at

Abstract Quantitative and qualitative models and reasoning methods for diagnosis are able to cover a wide range of different properties of a system. Both groups of methods have advantages and drawbacks in respect to fault diagnosis. In this chapter we propose a framework which combines methods of both groups to a combined diagnosis engine in order to improve the overall quality of diagnosis. Moreover, we present the different methods based on a running example of an autonomous mobile robot. Furthermore, we discuss the problems and research topics which arise from such a fusion of diverse methods. Finally, we explain how actively gathered observation are able to further improve the quality of diagnosis of complex systems.

Keywords Model-based diagnosis, Robustness, Fault-tolerance, Embedded systems

15.1 Introduction

Model-based diagnosis has been successfully applied to automated fault detection and localization in a wide range of different domains. Basically, diagnosis represents methods which are capable to detect a wrong behavior of a system and to find the root cause for that behavior. Applications for diagnosis comprise digital circuits, mobile robots, software development, automotive industry and space-probes [1–6].

From its beginning model-based diagnosis has been developed in several directions. These developments have been driven by the different characteristics of the application areas and their corresponding definition and use of models. A coarse partitioning of model is to distinguish qualitative

N. Martínez Madrid, R.E.D. Seepold (eds.), *Intelligent Technical Systems*, Lecture Notes in Electrical Engineering 38, DOI 978-1-4020-9823-9_15,

and quantitative models. Qualitative models are models that relate variables with finite value domains. Such models are based on an abstract logical description of the desired behavior of the system. The diagnosis process itself is done by logical reasoning [7] Examples for the use of qualitative models for diagnosis can be found in [8]. Quantitative models, on the other side, capture physical entities of a system in terms of real valued variables and the models relate them through differential-, difference- and algebraic equations. Continuous filtering of observed values [9–10] and the so called fault detection and identification (FDI) methods [11–12] solve the diagnosis task for this model class.

Both groups of methods cover different aspects of a system and have their advantages and drawbacks. Furthermore, the different methods provide diagnosis on a wide range of temporal, semantic and spatial granularity. This means that these diagnosis methods provide diagnosis at different frame-rates, as quantitative or qualitative information or about different parts of the system.

In general, diagnosis methods try to find an explanation for the current behavior of a system based on observations and models of the system. The behavior can be either nominal or faulty. Usually, the diagnosis process reports all diagnoses (explanations) that are consistent with the observations and the models. It is possible that this set of diagnoses contains misleading diagnoses or diagnoses that only roughly explain the observations of a faulty behavior.

Usually quantitative and qualitative models cover different aspects of the system. We believe that a combination of the outcome of these different models will improve the quality of the overall diagnosis. Moreover, it will reduce the number of diagnoses that are too general to be useful or that are simply wrong. In this chapter we propose a framework which combines the output of a number of qualitative and quantitative models in order to improve the quality of the overall diagnosis. The fusion of such different information is far away from being trivial and raises a number of interesting research questions.

Moreover, in general all these diagnosis methods use the observations the system provides during execution. There are no active methods in order to gather additional information which can help to improve the quality of the diagnosis. Therefore, we discuss the issue of active observations where the diagnosis system actively tries to gather additional useful observations. Obviously, such active observations require appropriate reasoning and planning capabilities.

In the remainder of this chapter we will first present some related work and will introduce a running example. An autonomous delivery robot will serve as an example for a complex system throughout the chapter. Using this example we will present the properties of quantitative and qualitative models in more details. In a following section we will present the proposed combined diagnosis framework and will discuss open questions that are raised by the framework. Moreover, we will briefly present and discuss the idea of active observations. Finally, will draw some conclusions about the ideas presented in this chapter.

15.2 Related Research

The Livingstone architecture proposed by Muscettola and colleagues [6] was used by the space probe *Deep Space One* to detect failures in its hardware and to recover from them. The fault detection and recovery are based on model-based reasoning.

In work Dearden and Clancy [13] and Verma et al. [14] particle filter techniques were used to estimate the state of the robot and its environment. These estimations together with a model of the robot were used to detect faults. The most probable state is derived from unreliable measurements. The advantage of this approach is that it is able to handle non-Gaussian uncertainties of the robot's sensing and acting as well as uncertainties in its environment. Other approaches which are based on Kalman-filter are only able to account Gaussian uncertainties.

Model-based diagnosis also has been successfully applied for fault detection and localization in digital circuits, in car electronics and for software debugging of VHDL programs [1].

Struss presented an approach for knowledge compilation for a diagnosis system [5]. The model and the reasoning process for the diagnosis of parts of the electronics of a car were condensed to a decision tree. This reduction of the demand for resources allowed them to apply the diagnosis system in an ordinary control unit for cars.

Roos described an algorithm which allows a group of diagnosis agents to negotiate a decision about a global diagnosis [15]. The agents had only a local view of the system. The main issue of this work was to minimize the communication overhead needed for a global diagnosis.

In the work of de Kleer a method was presented which is able to decide which measurement should be done next in order to gather maximum information for the diagnosis [16]. The development of this method was mainly driven by the fact that not all measurement contains the same useful information all the time. Furthermore, some measurements are more costly or harder to obtain than others. Therefore, one likes to minimize the use of such measurements if possible.

15.3 Running Example

Figure 15.1 shows the autonomous delivery robot Wonko. The robot comprises a differential-drive robot platform, a laser range-finder, a GPS receiver, an inertial measurement unit (IMU) and an Intel-based central PC running Linux. The robot is able to autonomously navigate around the campus of the university in order to deliver goods.

Fig. 15.1 The autonomous delivery robot Wonko

The central PC runs a three-layer control software. It comprises a hardware layer, a continuous layer and an abstract layer. The hardware layer is responsible for the interface to all hardware components. The continuous layer performs any continuous information processing like processing of the sensor inputs or reactive control. The abstract layer hosts the deliberative control in form of planning and reasoning module. The continuous layer is able to provide two different kinds of observations about the robot itself and its environment. The layer either provides processed information like the global position and orientation of the robot and the position of recognized objects or it provides unprocessed sensor data like the acceleration or turning rate measured by the IMU.

The presented robot platform will serve as an example for the diagnosis throughout the remainder of the chapter.

15.4 Quantitative Modeling

Figure 15.2 shows the arrangement of one axis of the drive of our robot. It comprises a wheel, a gearbox, a dc-motor and a wheel-encoder. Furthermore, it shows the inputs, e.g., commanded velocity, and the outputs, e.g., drawn current and actual angular velocity, we are able to observe.

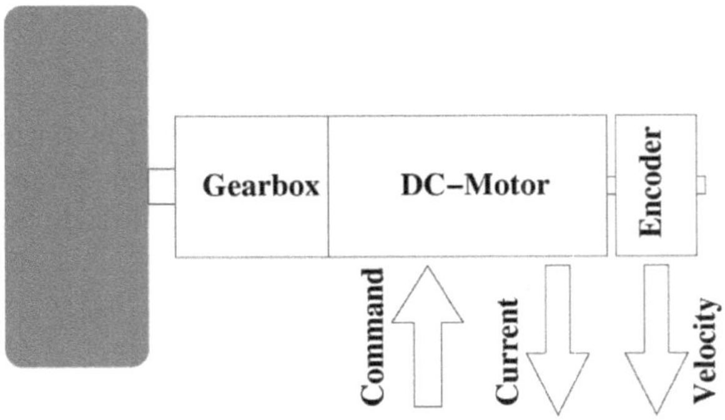

Fig. 15.2 The arrangement, the inputs and outputs of a single axis of the robot drive

In order to be able to detect and localize a fault in such an electromechanical system we have to model the desired behavior of the system. In general, an abstract qualitative model is not sufficient to model all aspects of such systems which are able to provide useful information for diagnosis. Therefore, we have to model the behavior of the system in all nominal and faulty operation modes which we want to detect. An approved method to model the behavior of such systems is a probabilistic hybrid automata. A detailed overview on this technique can be found in work of Hofbaur [17]. A probabilistic hybrid automata is an automata which comprises all nominal and faulty operation modes of the system as discrete states. Also a node is added to cover all the unknown operation modes which are not modeled exactly. Moreover, the automata comprises transitions between the modes and probabilities for their occurrence. Furthermore, a model of its dynamics is attached to each discrete state, i.e., difference or differential equations. This model describes how the continuous state vector of the system evolves in that particular mode. Figure 15.3 depicts a simple probabilistic hybrid automata which models the drive of the example robot comprising two of the axis depict in F Fig. 15.2. The figure shows the automata with the five nodes for nominal, faulty and unknown modes and their transitions. The probabilities for each transition were omitted for readability.

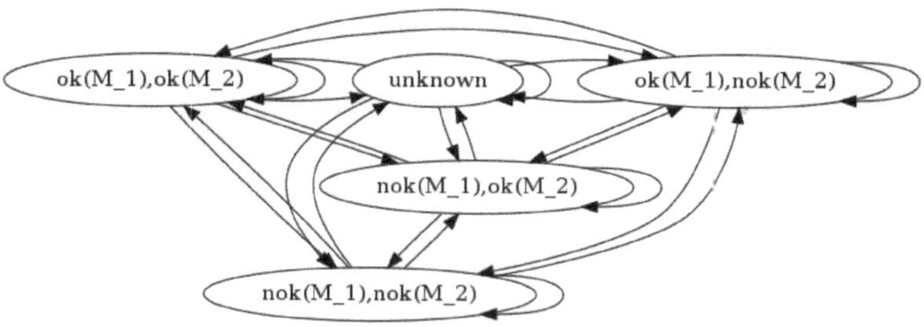

Fig. 15.3 A simple hybrid automata modeling the drive of the delivery robot

Equation (15.1) shows the differential equation for the nominal mode. Equation (15.2) shows the differential equation for the mode where the right motor fails. The equations of the unknown mode are empty.

$$\begin{bmatrix} \dot{\omega}_r \\ \dot{\omega}_l \end{bmatrix} = \begin{bmatrix} 1/\tau & 0 \\ 0 & 1/\tau \end{bmatrix} \begin{bmatrix} \omega_r \\ \omega_l \end{bmatrix} + \begin{bmatrix} 1 & 0 \\ 0 & 1 \end{bmatrix} \begin{bmatrix} u_r \\ u_l \end{bmatrix} + W \tag{15.1}$$

$$\begin{bmatrix} \dot{\omega}_r \\ \dot{\omega}_l \end{bmatrix} = \begin{bmatrix} 1/\tau & 0 \\ 0 & 1/\tau \end{bmatrix} \begin{bmatrix} \omega_r \\ \omega_l \end{bmatrix} + \begin{bmatrix} 0 & 0 \\ 0 & 1 \end{bmatrix} \begin{bmatrix} u_r \\ u_l \end{bmatrix} + W \tag{15.2}$$

The task of detection and localization of a fault in this case is equivalent to find the most probable operation mode based on observations of the system. The state estimation can be performed by multi-hypothesis tracking. The approach tracks all possible mode sequences over time and estimates how likely these sequences are with respect to the observed input and output sequence. That mode sequence with the highest likelihood captures most probable the true mode sequence.

15.5 Qualitative Modeling

In contrast to the quantitative modeling schema, qualitative models use an abstract logic-based model of the behavior of a system in order to be able to detect and localize faults. One advantage of this technique is that only the correct behavior of the system has to be modeled. The principles were originally developed by Reiter [7]. The basic idea is to use an abstract model, i.e., horn clauses for efficiency reasons, of the correct behavior of the system and some current observations, i.e., literals, of the system. If the prediction of the model differs from the observation a contradiction occurred and we have detected a fault. The check for such contradictions can be performed easily by logical inference. So far we only know that a fault occurred but we do not know the root cause of a fault. In general it is not obvious which faulty component is responsible for the observation of a faulty behavior. In order to solve this problem efficiently Reiter proposed the hitting-set algorithm. The idea is to systematically take back assumptions about functioning components until the contradiction is removed. Therefore, the removed assumptions about correct components are an explanation for the faulty behavior. These explanations are called a diagnosis. More details about model-based diagnosis can be found in book of Hamscher [8].

In the work of Friedrich et al. [1] a more convincing modeling schema was presented. The schema eases the modeling of larger systems. The idea is that in a first step only the correct behavior of smaller components is modeled, e.g., gates in integrated circuits. In a second step a structural description of the connections and interactions of the components is provided. The advantage is that the behavioral description of the components can be easily reused for other systems and only the structural description has to be adapted to the new system. In the remainder of the chapter we follow this schema for the qualitative modeling. We continue with an example which depicts the quantitative modeling.

The behavior of the differential-drive of the robot of Fig. 15.1 can be qualitatively modeled with the following simple clauses:

1. *¬AB(MOTOR_1) ∧¬AB(ENCODER_1) ∧turn(MOTOR_1)→ ok(TICKS_MOTOR_1)*
2. *¬AB(MOTOR_2) ∧¬AB(ENCODER_2) ∧turn(MOTOR_2)→ok(TICKS_MOTOR_2)*
3. *¬AB(MOTOR_1) ∧turn(MOTOR_1)→ok(CURRENT_MOTOR_1)*
4. *¬AB(MOTOR_2) ∧turn(MOTOR_2)→ok(CURRENT_MOTOR_2)*

Line 1 specifies that if the encoder and the motor of axis 1 works correct and we order the motor to turn we are able to observe encoder increments from axis 1. The predicate *¬AB(C)* denotes that a component *C* is not abnormal meaning that *C* works as expected. *ok(o)* denotes that the correct observation *o* has been made. Line 3 specifies that if the motor works correct and we order the motor to turn we can observe that the motor draws current.

Now we assume that the encoder of the motor of axis 1 fails and does not provide any encoder ticks anymore even if the motor turns correct. This observation can be expressed as follows:
turn(MOTOR_1) ∧turn(MOTOR_2) ∧¬ok(TICKS_MOTOR_1) ∧ok(TICKS_MOTOR_2)

If we assume that all components work correct, expressed by the clause *¬AB(MOTOR_1) ∧¬AB(ENCODER_1) ∧¬AB(MOTOR_2) ∧¬AB(ENCODER_2)* we have a contradiction between the outcome of the model and the observation. We can derive *ok(TICKS_MOTOR_1)* and *¬ok(TICKS_MOTOR_1)* at the same time.

This means that we have detected a fault. If we systematically remove assumptions about working components, i.e., change $\neg AB(C)$ to $AB(C)$, we are able to find one or more sets of removed assumptions which resolve the contradiction. These sets are the diagnosis of the system and describe the root cause of the detected fault. It has to be noted that the set comprising all components always forms a diagnosis. In the example the set *{AB(ENCODER_1)}* resolves the contradiction and is the true root cause of the fault. In general we are interested in diagnosis with minimal cardinality. Because of the fact that this approach tries to maintain consistency the technique is also called consistency-based.

15.6 Combine Quantitative and Qualitative Models

We presented in the previous sections different modeling and diagnosis paradigms. They differ mainly in the used modeling method (qualitative, quantitative or hybrid) and in their reasoning process (probabilistic state estimation, rule-based systems or logical inference). Furthermore, diagnosis methods can differ in their temporal and diagnosis granularity. Some models may deliver rough diagnosis on a fine temporal granularity for the whole system like there is a general fault in the control software. Other models may be capable to provide a much more specific diagnosis about a limited area of the system like the maximum possible angular velocity of a motor is suddenly limited to some value.

This shows that the different methods are capable to cope with different aspects of the system with different qualities. Moreover, in general diagnosis methods provide all explanation which are consistent with the model and the observations or which have a probability above a particular threshold. But all these explanations do not necessarily comprising the real root cause of the fault. It is similar to the situation when one brings its car to the garage because the motor makes a strange noise. The mechanic often provides the explanation that some components cause the noise. In general the true explanation is in the set but further investigation is necessary. Reconsider the example of the failed motor in Section 15.4. If we are unable to measure the drawn current both diagnosis *{AB(ENCODER_1)}* and *{AB(MOTOR_1)}* explain the observed behavior. But if we use a measurement of acceleration of the robot together with a quantitative model of the kinematics of the robot for all modes then we are able to reduce the number of diagnosis the correct one again.

In order to use the advantages of the different diagnosis methods while avoiding the drawbacks we propose to combine the different methods to improve the overall performance of the diagnosis. We believe that beside the better quality of the diagnosis with that approach we are able to handle much more complex systems comprising hardware and software.

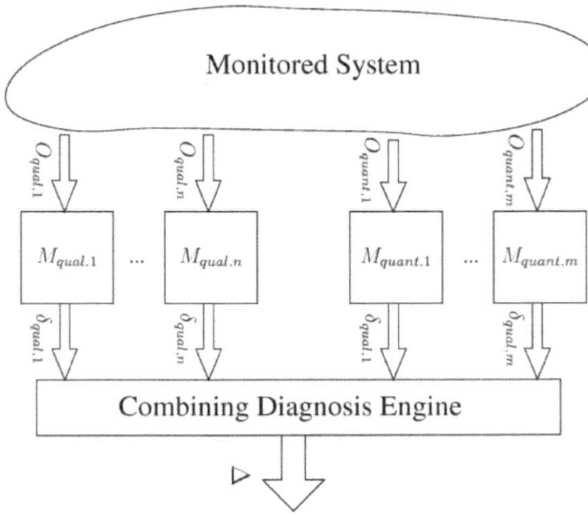

Fig. 15.4 The figure shows the framework for combining qualitative and quantitative models and diagnosis in order to improve the overall diagnosis Δ

Figure 15.4 depicts an overview of an architecture which combines different models for the diagnosis process. *M* denoted the qualitative and quantitative models and reasoning modules. *O* denotes the different observations of the system used as inputs for the different diagnosis processes. The observations may origin from different parts of the system and depict various aspects of the system like continue valued measurements or abstract literals.

δ denotes the outcome of these local diagnosis processes. The temporal and diagnosis granularity of these diagnoses can be very different. While one model may deliver a full set of components that may have caused the fault another module may just deliver an estimation about the operation mode of a component. Moreover, the diagnosis output of one model can be used as an input for another model. For instance a quantitative model may provide an input predicate for a qualitative model. All the different diagnoses δ represent a different view on different parts of the system. Moreover, they comprise different knowledge and opinions about the state of the system. Obviously, these different parts of knowledge will be sometime inconsistent and will provide different explanations for an observed faulty situation.

In order to increase the overall quality of the diagnosis we propose to combine the different local diagnosis results to a global one. Δ in Fig. 15.4 denoted the combined improved overall diagnosis of the system. The combination of such different models and reasoning methods will increase the quality of the resulting quality of the diagnosis but raises a number of open questions:

- **Different Temporal Granularity** The different models may use observations that come at a very wide range of frame-rates. A hearth-beat message from a component may be received at a frame-rate of about one Hz while other sensors may provide data at a frame-rate up to several hundreds of Hz. Due to this factor also the different reasoning engines may provide their results on such a wide range of frame-rates. Moreover, extensive logical reasoning or the temporal integration of data in the filtering and tracking methods may cause further delays of the diagnosis with respects to the time the observations were made. Therefore, appropriate methods are required in order to synchronize the local diagnosis to avoid contradictions or inconsistencies. There are very good approaches in sensor fusion which addresses the fusion of (uncertain) information provided at different frame-rates and sampling times [18].
- **Different Diagnosis Granularity** As described in the previous sections the different models and reasoning methods provides diagnosis on a wide range of semantic granularity. On one hand very general abstract diagnoses are provided. In general they are derived by logical inference from an abstract logic-based model. On the other hand filtering and tracking methods are able to provide information about a system in very fine resolution such as estimations about the continuous state vector of a system. The fusion of information on an abstract symbolic level is much easier. In order to be able to fuse continuous valued information at that level we have to perform symbol grounding for those values. Symbol grounding can be done in the easiest case by a simple threshold. The disadvantage of such methods is that one probably throws away useful information obtained by quantitative models. Therefore, methods for a direct fusion of quantitative and/or qualitative methods are needed.
- **Spatial Distribution** Another important issue is that different models may provide diagnosis about different parts of the system. One model may provide a diagnosis about the drive hardware of a robot while another model may provide a diagnosis about the control system of the robot. Due to the fact that often a fault in one component also causes a depending fault in an interacting component sometimes false alarms are raised. In order to improve the quality of the diagnosis we have to combine the output of different models. Therefore, the development of meta-models is necessary in order be able to do this information fusion. There exists work about distributed diagnosis but they use in general the same semantics in the output of the models. Therefore, new more general meta-models have to be developed.

- **Performance Issues** In general the presented diagnosis methods are expensive in terms of computational power and memory. This is a limiting factor if such intelligent methods should be deployed in small embedded systems. Anyway, subsets of these methods have already been deployed to embedded systems in cars for instance [5]. These applications loose some level of flexibility of the methods and compile their knowledge into simpler structures like decision trees. In the future models with less demand on computational resource have to be developed.

Despite the fact that qualitative and quantitative models for debugging make use of different modeling techniques both are used for identifying the root causes of a detected misbehavior. The root cause is either a state a system has to be in order to explain the observed behavior or a set of components that behave in an unexpected way in order to resolve the contradiction between the observed and the deduced behavior. Both representations have in common that they assign a mode, e.g. faulty or stuck-at, to a component. Although, the granularity of diagnosis components might be different in different models in practice it should be possible to come up with a mapping that allows for comparing the diagnosis outcome.

For the following procedure for combining different models we assume the existence of such a mapping. In particular we assume that we have 2 different models M_1 and M_2 where M_1 is a more abstract representation of the system's behavior than model M_2. For both models we assume that we have diagnosis components $COMP_1$ and $COMP_2$ respectively, which might be different. Furthermore, we assume the existence of a function Γ: $COMP_2 \mapsto COMP_1$, which maps a component of the more precise model to the more abstract one. Note that Γ needs not to be a bijective function and more than one component of M_2 is mapped to the same abstract component of M_1.

The combination of such models would improve diagnosis time if the diagnosis output of one model can be used to focus the diagnosis computation using the other model. Formally, the diagnosis problem is a tuple (*SD*,*OBS*,*COMP*) and we assume a procedure **DIAGS**, which returns the set of minimal diagnoses for the given diagnosis problem. *SD* is the system description which represents the behavior of the different components and it connections. *OBS* is the set of actual observations of the system. *COMP* is the set of the components of the system. Given these basic definitions, we can solve the problem of combining diagnosis model using the following algorithm where we assume two sets of observations OBS1 and OBS2 that can be used together with the two models in order to compute diagnoses:

1. Compute the set of diagnoses D = **DIAGS**($SD_1, OBS_1, COMP_1$) using the more abstract model M_1.
2. Compute the set of components that are used in at least one diagnosis, i.e. $COMP' = \{ C \mid C \in \Delta, \Delta \in D \}$
3. Compute the focus set for model M_2 such as follows: $COMP'' = \bigcup_{C \in COMP'} \Gamma^{-1}(C)$.
4. Compute the set of diagnoses D' = **DIAGS**($SD_2, OBS_2, COMP''$) and return D' as result.

The algorithm allows for a direct integration of two models. If $COMP''$ is a subset of $COMP_2$, then the algorithm allows for focusing on those components that pass the first diagnosis step. Hence, diagnosis time can be improved especially in cases where the second model would require a lot of computational resources when checking all components. The algorithm can be further improved when assuming that **DIAGS** only gives back the most probable diagnosis candidates.

Note that this algorithm cannot be used if the underlying assumptions do not hold. It is especially important that a mapping function Γ exists. Moreover, there might be further improvements when not only considering mappings between components but between fault models, which are usually handled by introducing fault modes together with an assigned fault behavior. In this case the component together with the fault modes has to be reduced using the outcome of a more abstract model.

15.7 Active Observations

In general the observations of the systems are caused by the ordinary operation of the system. The system is performing its task and provides observations about itself. Usually, the diagnosis module has no control about which actions the system is performing and therefore what observations are available at that moment. Furthermore, it might happen that exactly these observations do not contain the necessary information which is needed to do the best possible diagnosis.

We assume the following situation. During the movement along a given path some component in the robot drive fails. Usually, the robot continues it started action until it takes another decision. The diagnosis modules only can rely on the observation produced by that current action. In some case these observations might be not sufficient in order to locate the failed component. Therefore, in some situations it will be desirable that the diagnosis and monitoring system is able initiate actions in order to gather more useful information. In the above example such an action can be a special motion pattern or profile for the motors. Such additional observation can help to reject not plausible diagnosis.

The application of such active observation request for additional planning and reasoning capabilities because the diagnosis module has to derive which additional information is useful in order to improve the diagnosis and which action will provide the required information. Furthermore, the system has to take care that these additional actions do not endanger the system or the environment even in the case of an occurred fault. For this purpose additional models and planning strategies have to be developed.

15.8 Conclusion

In this chapter we proposed the combination of quantitative and qualitative models and reasoning methods in order to improve the diagnosis capabilities for complex systems comprising hardware and software. Moreover, we presented the properties of quantitative and qualitative diagnosis methods. Based on a running example of an autonomous mobile robot we motivate that such a combination is valuable and how it can be realized. Moreover, we pointed out potential problems and research topics related to the combined diagnosis. These problems mainly concern the useful fusion of temporal, semantic and spatial different information. Furthermore, we introduce the idea of active observations which are a way to actively gather additional information for the improvement of the quality of diagnosis. We believe that the proposed framework will influence the achievable quality of supervision. Our future work will be focused in a first step on the fusion of the different diagnosis information.

References

1. Friedrich G., Stumptner M., and Wotawa F., Model-based diagnosis of hardware designs. Artificial Intelligence, 111(2):3–39 (1999).
2. Hofbaur M., Köb J., Steinbauer G., and Wotawa F., Improving robustness of mobile robots using model-based reasoning. Journal of Intelligence and Robotic Systems, 48(1):37–54 (2007).
3. Steinbauer G. and Wotawa F., Detecting and locating faults in the control software of autonomous mobile robots. In 19th International Joint Conference on Artificial Intelligence (IJCAI-05), pp. 1742–1743 (2005).
4. Köb D. and Wotawa F., Introducing alias information into model-based debugging. In Ramon Lopez de Mantaras and Lorenza Saitta, editors, Proceedings of the 16th Eureopean Conference on Artificial Intelligence, ECAI'2004, pp. 833–837 (2004). IOS Press.

5. Struss P. and Price C., Model-based systems in the automotive industry. AI Magazine, 24(4):17–34 (2004).
6. Muscettola N., Nayak P., Pell B., and Williams B., Remote agent: To boldly go where no AI system has gone before. Artificial Intelligence, 103(1–2):5–48 (1998).
7. Reiter R., A theory of diagnosis from first principles. Artificial Intelligence, 32(1):57–95 (1987).
8. Hamscher W., Console L., and de Kleer J., Readings in Model-Based Diagnosis. Morgan Kaufmann, San Mate. (1992).
9. Kalman R., A new approach to linear filtering and prediction problems. ASME Transactions, Journal of Basic Engineering, 82:35–50 (1960).
10. Anderson B. and Moore J., Optimal Filtering. Information and System Sciences Series. Prentice Hall, Englewood Cliffs (1979).
11. Isermann R., Supervision, fault-detection and fault-diagnosis methods - an introduction. Control Engineering Practice, 5(5):639–652 (1997).
12. Chen J. and Patton R., Robust Model-Based Fault Diagnosis for Dynamic Systems. Kluwer, Dordrecht (1999).
13. Dearden R. and Clancy D., Particle filters for real-time fault detection in planetary rovers. In Proceedings of the 13th International Workshop on Principles of Diagnosis, pages 1 – 6 (2002).
14. Verma V., Gordon G., Simmons R., and Thrun S., Real-time fault diagnosis. IEEE Robotics & Automation Magazine, 11(2):56 – 66 (2004).
15. Roos N., ten Teije A., Bos A., and Witteveen C., Multi-agent diagnosis with spatially distributed knowledge. In Proceedings of the Belgium-Netherlands Artificial Intelligence Conference (BNAIC), pp 275–282 (2002).
16. de Kleer J., Getting the probabilities right for measurement selection. In 17th InternationalWorkshop on Principles of Diagnosis (DX-06), pp 141–146 (2006).
17. Hofbaur M., Hybrid Estimation of Complex Systems, volume 319 of Lecture Notes in Control and Information Sciences. Springer Verlag, New York (2005).
18. Drolet L., Michaud F., and Cote J., Adaptable sensor fusion using multiple kalman filters. In Proceedings IEEE/RSJ International Conference on Intelligent Robots and Systems (IROS). (2000).

Part V

Intelligent Engineering

Chapter 16

Object Memory Management for Constrained Devices with Heterogeneous Memories

Kevin Marquet and Gilles Grimaud
CNRS/INRIA/Univ. Lille 1, France, Kevin.Marquet@lifl.fr

Abstract Small devices have a specific hardware configuration. In particular, they usually include several types of memories (typically internal and external RAM, EEPROM, Flash), different in quantities and properties. For instance, their access times can be very different. This is an issue for object-oriented solutions such as Java virtual machines which have to perform automatic data reclamation. In this paper, we firstly present results showing that the memory manager (especially the garbage collector) must be adapted to the type of memory it is in charge of. Then, we propose a flexible memory management solution that addresses this issue by assigning a different memory manager to each memory. Each manager can use the allocation and garbage collection strategy adapted to the physical properties of the memory it is in charge of. In order to handle interactions between memory managers during allocations and garbage collections, we use special components in charge of synchronizing managers. Thereby, our solution brings the benefits of automatic data reclamation to devices with heterogeneous memory spaces.

Keywords Memory management, Garbage collection, Operating systems, Virtual machines

16.1 Introduction

In order to obtain a compromise between their different physical capabilities, reading and writing speeds, density ($bits \triangleleft mm$ - of silicium) persistence and security of data, small devices include several types of physical memories. Typically, several types of EEPROM and RAM can be included in addition to the read-only memory. Thus, systems embedded on such devices need to manage several memory spaces and to deal with their

N. Martínez Madrid, R.E.D. Seepold (eds.), *Intelligent Technical Systems*, Lecture Notes in Electrical Engineering 38, DOI 978-1-4020-9823-9_16,

physical properties. Moreover, memories are present in uncommon proportions. A device of the range of a smart cards usually include few kilobytes of RAM, dozens of kilobytes of EEPROM and/or Flash memory, and hundreds of kilobytes of ROM. These characteristics imply the use of a memory management model able to manage several memories, to put data in those memories efficiently, and flexible enough to be usable on various types of devices. Actually, many embedded systems require applications programmers to choose memory in which allocating data and to release unused resources. Such solutions prevent the portability and the reuse of applications because the memory configuration is not the same on different devices, in capacity as well as in physical properties. Moreover, application programmers must manage the memory themselves, and are therefore required to have deep knowledge of the hardware functioning. The use of object-oriented solutions is an effort to increase the portability of applications for smart devices, in particular thanks to the use of garbage collectors. In this context, we propose a memory management model in which each memory space is handled by a particular memory manager. Our solution provides several benefits:

- the use of automatic data reclamation algorithms provide a certain safety; we intend to bring this safety to devices with several memory spaces;
- associating one specific manager to a memory space allows to manage differently each memory according to its properties;
- the programmer of applications must not necessarily define the location of objects.

This chapter is organized as follows. Section 16.2 presents past research works on both memory-management in embedded systems and solutions to manage several memories. Section 16.3 details our solution to allow garbage collection on devices with different memories. Section 16.4 gives experimental results. At last, conclusions and perspectives on this work are given in Section 16.5.

16.2 Multiple Memories Management

Some automatic data collection algorithms are particularly well-suited for embedded use. This section firstly describes these types of garbage collectors; this will ease the comprehension of the solution proposed in Section 16.3. Then, remaining problems are presented.

16.2.1 Garbage Collectors for Embedded Systems

Previous works have shown the effectiveness of certain types of garbage collectors. These garbage collectors have different functionning and properties. Although a large number of collectors exists, we consider and describe only a few of them which are particularly well-suited for embedded use.

Since it does not move objects during collection, the mark and sweep collector (M&S) [1] performs few memory writings. Thereby, it is a good means to collect data in memories slow in writing, like EEPROM. This collector only performs two phases: **mark** live objects, and **collect** unused data

Semi-space copying collectors [2] are fast, typically operating as follows. First, it **traces** live objects, move them to the unused semi-space. Second, it scans memory and **updates** references.

The cldc Hotspot virtual machine uses a 2-generational copying collector [5], which performs minor collections during which live objects are copied from the nursery to the bottom part of the heap:

Mark: traces all live objects of the nursery and marks them as such. **Collect**: copies all live objects of the nursery in the bottom of the heap. **Update**: updates all references to match the new location of objects previously located in the nursery.

Compacting collectors [3] have small memory overhead and low power consumption [4]. They typically operate as follows: **Mark**: traverses all live objects and mark them as such. **Prepare**: scans memory, computes the new location of objects, which is stored into an extra word of the object. **Modify**: scans the memory, replaces all references with the new addresses. **Collect**: moves each live object to its new location. As for semi-space collectors, only live objects are moved, thereby collecting others implicitly.

Several versions of this algorithm exist. It is possible to build a table, in order to store only references of live objects and obtain good cache effects: **Mark**: traverses all live objects and marks them as such. **Prepare**: scans memory and stores the address of each live objects into a table of references (addresses increasingly ordered). Version 1: each entry of the table contains one word. Version 2: the address is stored in the first word of a two-word entry. **Modify**: (Version 1) dereferences all references towards the table. **Collect**: moves each live object to its new location. Version 1: this new address erases the old location in the table. Version 2: this new address is stored in the second word of the entry.

Update: update each reference in memory to point to the new location.

Version 1: this new location is the one pointed by the reference in the table. Version 2: this reference is found in the second word of the entry.

16.2.2 Problems with Several Memories

Many garbage collection techniques have been developed along the past decades [6]. A recent work [4] presents garbage collectors operating well on small devices. All these techniques focus on optimizing garbage collectors but give no means to manage together all memories of a given device.

The different memories of a small device can be seen as different regions. Managing memory with regions has been the focus of a number of works. A number of these works requires regions to be explicitly named at allocation and removal [7]. Others [8] use annotations. We are not concerned by these solutions as we want memory to be reclaimed automatically. Other work proposes a partitioning of the memory based on a static analysis of data type [9]. These models suppose that the size of regions can be changed. In our case, it is not possible since regions are physical memories.

Generational garbage collection techniques [10,11] store objects in regions on the basis of the age of objects. This criterion is accurate for traditional virtual machines but cannot be the only one in embedded systems. Indeed, if the memory space used for old objects is not efficient (for instance the amount of available EEPROM or Flash memory is greater than other memories), an old object often modified will slow down the execution.

The Java Card [12] memory management scheme is to allocate all objects in EEPROM. The programmer can use a special library to allocate in RAM. Several problems arise with this solution. Firstly, it is not portable because it only makes sense with one specific hardware architecture (the smart cards one). This solution is not applicable to devices that use different types and/or sizes of memories. Secondly, the use of libraries, which is mandatory in most cases, is problematic: as libraries can be used in different manners, objects allocated in those libraries will not be accurate for all applications.

Each type of physical memory has its own properties, in terms of security, life time, size and cost (RAM is expensive and takes an important place on the device). Most of all, reading and writing access times can be very different. For instance, writing in EEPROM is thousands times slower than reading it. Such differences require managing each memory in an adapted way. In order to illustrate the impact of the type of memory, we measured [13] the time spent in memory access of different applications using different collectors. Results showed that the best collector for a given application is not always the same, regarding to the access speed of the underlying memory.

In order to address these issues, we propose a solution that makes it possible to specialize the memories management regarding to the device and the applications.

16.3 Proposal

Our proposal is based on these two ideas:

- memories have different properties; thereby, let each one be managed by a specific memory manager;
- the placement of data in memories is a transversal concern.

Our solution is to assign one type of memory manager (including a specific type of collector) to each memory, in order to manage efficiently their properties. This is illustrated by Fig. 16.1: instead of designing a memory management system dedicated to specific hardware architecture, we have designed a framework allowing each memory space to be managed by a memory manager.

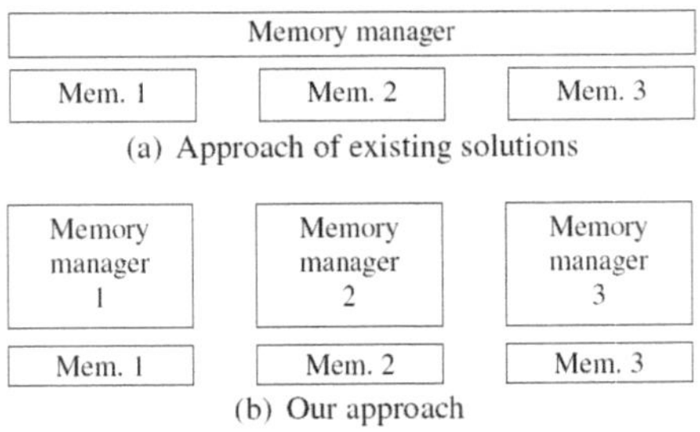

Fig. 16.1 Comparison of traditional solutions with our approach

16.3.1 Overview

Basically, two actions can be performed on a memory space: allocations in the memory space, and collection of the space. These tasks are performed by the memory managers and can thereby be different for each memory.

However, some tasks cannot be performed by a specific manager. First, the choice of placing objects in one memory or another. Secondly, the marking of objects (as this task is transverse), it cannot be performed by one specific manager. At last, references can point from one memory to another. As objects can move (compacting collectors for instance), updating references must be considered.

That is why two special components are responsible for handling those tasks. The Placer is responsible for choosing the location of data. The Collector is responsible for collection phases. It performs marking phases, and initiates the collection phases so that references. These components are the only parts of the code managing memories visible from the rest of the system. More specifically, the Placer provides an allocator, and the Collector provides a collection function. These two components use services provided by memory managers, accessible through a dedicated API. We show in the following that any automatic memory management system can be implemented through this API. Figure 16.2 shows the general architecture of this system. It is detailed in the following.

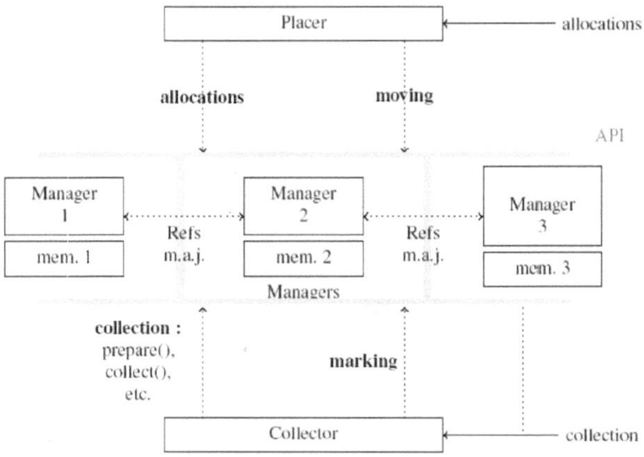

Fig. 16.2 General architecture.

16.3.2 A Distributed Marking Algorithm

The marking phase is a problem in the case of multi-heaps management. Indeed, during this phase, writings can be performed in the different memory spaces, and inter-partitions references must be followed. However, we want each manager to be responsible for writings in its partition. Therefore, it is necessary to use a marking algorithm that takes into account this constraint. To this goal, we inspired from so-called distributed algorithms [14], initially invented to collect data in multiprocessor environment. Different types of distributed algorithms exist, based on Dijsktra [15] or Ali [16] works. However, they all require additional memory space, which we do not want. In addition, they generally require data heap organization to be either the same in all heaps, or well known (for centralized solutions).

The standard, efficient and simple way to mark live objects is to use a stack [14]. Roots references are pushed; then, while the stack is not empty, a reference is unstacked and the pointed-to object is scanned. If they are not marked, its children are marked and their references pushed on the stack. In this manner, all live objects are marked at the end of the phase.

On this basis, we designed a distributed algorithm able to mark objects in different heaps. It uses services provided by memory managers so that writings are performed by them. This algorithm allows marking live objects in different heaps; in addition, it allows each manager to mark objects in its specific manner. Figure 16.3b shows the standard algorithm as well as the distributed algorithm we invented. The principle is therefore similar but the covering loop has changed: iterations are done on the managers while at least one of them has marked a new object (whose children must be scanned to eventually be marked). At each iteration, the current manager is notified (call to mark_heap()) that it has to handle marked objects of its heap (but whose childs have not been reached). When an object is reached, the manager of the partition in which it is stored is asked to mark it (call to mark_object()).

Figure 16.3b shows a typical implementation of mark_heap() and mark_object() functions, with the use of a marking stack. In the case of a semi-space copying collector, a simple implementation would be to mark each object and move it to the unused semi-space.

16.3.3 Allocation Management

When an allocation must be performed (i.e. execution of a new, byte code), the allocator of the Placer is called. This allocator is responsible for determining in which region the new object should be allocated. Once the selection of the region is made, the allocator of the memory manager in charge of the selected memory space is called. This requires that each memory manager provides an allocator. Managing the placement of objects directly at their creation allows to avoid the useless displacement of objects and can avoid to fill up too quickly a memory present in little quantity. Moreover, it allows allocating judiciously certain objects. For instance, an execution stack should be allocated in a high speed memory as it is often read and modified. The placement of data is detailed in Section 16.3.5.

16.3.4 Garbage Collection

From the typical types of collector presented in Section 16.2, it is possible to retrieve common points for the functioning of garbage collectors. Firstly, all these collectors require live objects to be marked. Then, they all perform a collection phase. For compacting and copying collectors, this phase consists in moving all live objects whereas, for the mark-sweep collector, it consists in reclaiming unused memory. At last, collectors that move objects need to update the references contained within objects. In order to perform this phase, the table-based version of the compacting collector needs to perform a phase where all references are updated to point to an indirection table.

```
mark() =
  mark_stack = empty
  for R in Roots
    mark_bit(R) = marked
    push(R, mark_stack)
  mark_heap()

  while mark_stack ≠ empty
    N = pop(mark_stack)
    for M in Children(N)
      if mark_bit(M) == unmarked
        mark_bit(M) = marked
        push(M, mark_stack)
```

(a) Standard algorithm

```
mark() =
  done = false
  for MM in Managers
    MM- >prepareToMark()

  for R in Roots
    for O in Children(R)
      MM = getManagerOf(O)
      MM- >mark_object(O)
  while not done
    done = true
    for MM in Managers
      done = not MM- >mark_heap()


boolean mark_heap() =
  while MM- >mark_stack ≠ empty
    N = pop(MM- >mark_stack)
    for O in Children(N)
      OM = getManagerOf(O)
        OM- >mark_object(O)
        marked = true
  return marked


boolean mark_object(Object O) =
  if mark_bit(O) = unmarked
    push(O, MM->mark_stack)
```

(b) Distributed algorithm

Fig. 16.3 Comparison of standard and distributed algorithms

16.3.4.1 Interactions Between Memory Managers

When a collection is performed, objects can move, leading to invalid references in other memory spaces. As a consequence, references in all memory spaces have to be updated. Two solutions can be used. Firstly, to implement write or read barriers [17] using a card marking mechanism [18]. Secondly, to scan spaces when needed in order to update the references they contain. Both the solutions are applicable with our proposal. Since write barriers (and a fortiori read barriers) are costly in time or in space, we choose the second one.

We think that moving objects from one region to another is a capability that must be provided, either to move objects in a more adapted region (see Section 16.3.5), or to free a memory space. This capability is used by generational garbage collectors. However, the location of objects during garbage collection cannot be evaluated by the memory managers since they have no knowledge concerning other spaces, other managers and their properties. Our Placer is responsible for finding the right location of an object. Given the role of different memory managers, Placer and Collector, we have defined a way to make them cooperate. When a manager needs to collect data in its memory space, it notifies the Collector. The latter triggers several phases successively which will allow to collect data and handle side effects of the collection. From our observations concerning existing garbage collectors (Section 16.2.1), the collection of a space (whose manager denoted by M) consists in four phases:

Prepare: During this phase, M performs preliminary operations. For example, compacting collectors build a references table if they need so.

Modify: In each memory space, references are modified if needed. For instance, the compacting collector variant makes all references of all regions to point to an indirection table during this step.

Collect: The collection of objects is made during this phase. For each live object, the memory-manager task the Placer whether the object must move.

Update: In each memory space, the references are updated if needed (if the collector has moved objects).

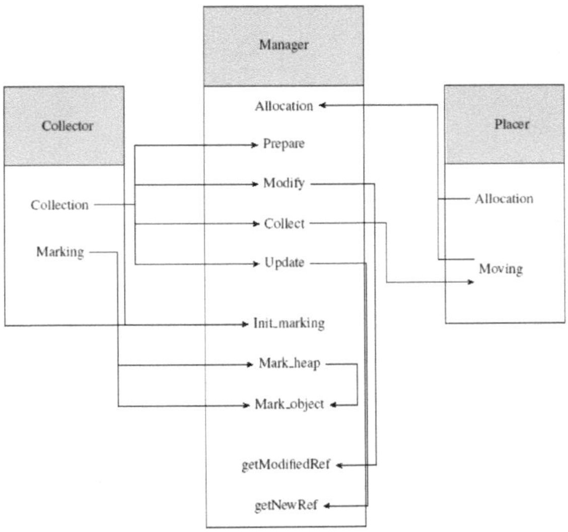

Fig. 16.4 Components of the overall system and links between services

Each memory manager must provide the four functions that perform these steps. These functions are part of the API that a manager must implement in order to be part of the framework we have defined. When a garbage collection of a region is needed, the Collector successively calls these functions, in order to complete the collection. Steps 1 and 3 are only performed by the manager that needs to collect its data, whereas steps 2 and 4 allow all managers to keep references up-to-date. These steps are initiated by the Collector; this ensures a good synchronization between the managers.

Figure 16.4 presents all components involved in our proposal and the links between the different services. The services are provided in our implementation through C function pointers.

Figure 16.5 illustrates the proposal. In this scenario, three memory managers exist (mm1, mm2, mm3). mm1 is a semi-space collector that performs a separated marking phase. After the marking phase, the Collector notifies mm1 to collect. Then, it asks successively mm2, mm3, and mm1 to update references.

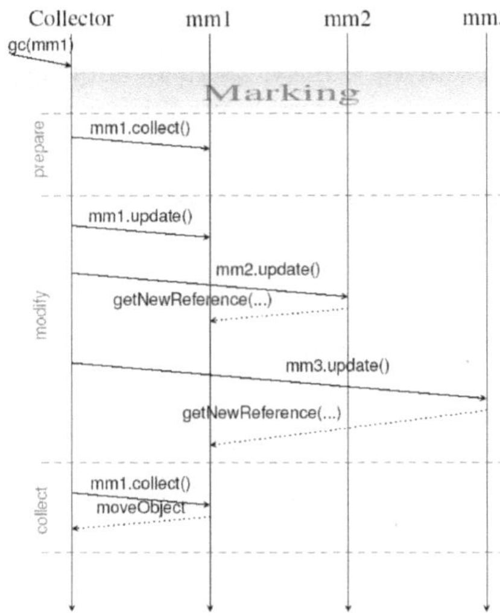

Fig. 16.5 A scenario where mm1 performs a collection

16.3.5 Placing Objects

It is the role of the Placer to place objects at allocation. The placement scheme can be defined according to the specifics of the memories as well as those of the applications. A generational model can easily be set up as it is enough to move, during a collection, all live objects from the nursery to the main memory space. The Placer for Java Card is very simple: it allocates in EEPROM or RAM. However, with the new generations of Java Card, more types of memory will be addressable (several types of Flash memory), and in larger amount. Our solution only requires changing the Placer to work with this new design. This reasoning is valid with any other memory configuration.

Thus, another part of our work aims at replacing the C code of the Placer with a language dedicated to the aspect placement of objects. We have designed a Domain-Specific Language which can be used to write placement policies. Such a policy uses some placement criteria (e.g. size, object type, allocation site, etc.) to express the placement of objects. The policy is compiled to generate the part of the Placer in charge of the placement decision.

16.4 Experimental Results

In order to estimate the proposed solution, we measured the execution speed, the sizes of code and data, and the flexibility of the architecture.

16.4.1 Efficiency

We performed measurements on the following benchmarks: dhrystone (evaluate general performances of processors); check (test properties of the virtual machine); raytracer (compute a 3D scene rendering); crypt (encrypt and decrypt a file); moldyn (simulate interactions between molecules). In order to evaluate the cost of our proposal, we now compare execution times of the managers within and outside the proposed architecture. For each type of collector, we measured the cost of the architecture, in percentage of the execution time outside the model. In this execution of reference, no function pointers are used, and variables are accessed directly rather than through C structures. Table 16.1 presents the cost of using our distributed marking algorithm.

For each collector, the average cost as well as the maximum and minimum costs is given. These measurements were obtained on an Intel Pentium 4, 3 GHz, 2MB cache, with a Linux system. However, marking is only a part of the problem, and Table 16.2 presents results for the total time spent performing collections.

The same experiments were conducted on a system corresponding to targeted platforms: processor ARM 9 200 MHz, 8 KB cache. Results are given in Table 16.3 and corroborate previous experiments. This shows that results illustrate algorithmic characteristics and not hardware specifics.

Experiments show that the impact of the proposal on performances is about +5% for the total collection time. This number is not neglect able but is low in reference with the time spent by a system collecting data. Indeed, the time spent in data collection is below 15% for reasonable memory needs and/ or collector choice [19,20]. In this case, a rise of 5% of the time spent in data collection means an overall slowdown of less than 1%. By the way, the flexibility of our solution allows improving performances by adapting memory management to applications, hardware, and software partitioning.

Table 16.1 Cost of the distributed marking algorithm

	Copying	Compacting	Compacting v1	Mark and sweep
Average	+9.1%	+7.7%	+8.7%	+8.8%
Max	+10.1%	+8%	+13.3%	+10.8%
Min	+7%	+7.4%	+6.1%	+7.8%

Table 16.2 Cost of the proposal on collection time

	Copying	Compacting	Compacting v1	Mark and sweep
Average	+3.8%	+5.4%	+1.7%	+5.6%
Max	+4.9%	+5.7%	+3.1%	+9.8%
Min	+2.7%	+4.6%	+0.7%	+3.5%

Table 16.3 Performance impact on collection time and marking time (ARM9)

	Copying	Compacting	Compacting v1	Mark and sweep
Marking	+8.9% (max 9.1%)	+9.3% (max +9.8%)	+9.6% (max +10.2%)	+8.1% (+11.2%)
Collection	+4.8% (max 5%)	+3.3% (max 4.1%)	+2.8% (max 2.8%)	+3.9% (max 4.1%)

16.4.2 Sizes of Code and Data

These measurements show that the proposed architecture can be embedded in constrained devices: Marking (7.4 kB), Placer (0.2 kB), Collector (0.2 kB), Other tools (2.3 kB), Copying collector (2.6 kB), Compacting collector (2.8 kB), Mark and sweep collector (2.8 kB), Compacting collector (version 1) (2.8 kB), Compacting collector (version 2) (3.5 kB). Between 2.5 and 3.5 kB per manager must be added. The 10 kB of tools are mainly due to marking management. Therefore, about 19 kB are required to embed three different memory managers, including a great part dedicated to marking management that has nothing to do with our proposal.

To these sizes, we add the space needed to store the structures describing memory managers: 12 read-only pointers, plus 4 writable pointers, leading to an overall 64 bytes if pointers are stored on four bytes.

16.4.3 Flexibility of the Architecture

The six memory management systems (allocator + collector) detailed in Section 16.2.1 has been implemented. They all work within the proposed framework, simply respecting the defined APIs. In addition, Salagnac [21] was able to insert its own memory manager in the framework. In fact, besides the references counting collector, which is anyway a problem (requires write barriers, does not detect cycles), we did not find any collector that could not work within the model. This shows that our solution is accurate and low restrictive.

16.5 Conclusion and Future Works

We have presented an innovative way to manage different memories on object-oriented systems. This solution allows managing differently the physical memories, in a way adapted to its properties. In this model, the placement of objects is a transverse preoccupation, which allows managing it more accurately by adapting it to the running application. We have briefly presented an implementation of this model. In this implementation, the memory manager of each memory space implements a set of functions allowing them to be part of a framework of managers. The managers of this framework are able to cooperate in order to allocate and collect data. Special components are used to choose the placement and synchronize them during a collection.

We intend to evaluate different placement schema, and see which ones are most efficient. It would also be interesting to be able to change the placement rules dynamically, for instance with the loading of new applications.

References

1. J. L. McCarthy. Recursive functions of symbolic expressions and their computation by machine. Commun. ACM, 3(4), 1960.
2. J. Cheney. A non recursive list compacting algorithm. Communications of the ACM, 13(11), 1970.
3. J. Cohen and A. Nicolau. Comparison of compacting algorithms for garbage collection. ACM Trans. Program. Lang. Syst., 5(4), 1983.
4. D. F. Bacon, P. Cheng, and D. Grove. Garbage collection for embedded systems. In EMSOFT Proceedings of the fourth ACM international conference on Embedded software, 2004.

5. Sun Microsystems Inc. The cldc hotspot implementation virtual machine. 2008.
6. P. R. Wilson. Uniprocessor garbage collection techniques. In IWMM ,Äô92 : Proceedings of the International Workshop on Memory Management, 1992.
7. D. Gay and A. Aiken. Memory management with explicit regions. In PLDI ,Äô98 : Proceedings of the ACM SIGPLAN 1998 conference on Programming language design and implementation. 1998.
8. D. Grossman, G. Morrisett, T. Jim, M. Hicks, Y. Wang, and J. Cheney. Region-based memory management in cyclone. In PLDI ,Äô02 : Proceedings of the ACM SIGPLAN 2002 Conference on Programming language design and implementation. 2002.
9. T. Harris. Early storage reclamation in a tracing garbage collector. SIGPLAN Not., 34(4), 1999.
10. H. Lieberman and C. E. Hewitt. A real-time garbage collector based on the lifetimes of objects. Commun. ACM, 26(6), 1983.
11. D. Ungar. Generation scavenging: A non-disruptive high performance storage reclamation algorithm. In Proceedings of the first ACM SIGSOFT/SIGPLAN software engineering symposium on Practical software development environments. 1984.
12. Sun Microsystems Inc. Java Card API 2.1 Specification.
13. K. Marquet and G. Grimaud. Garbage collection for tiny devices: A complexity study. In Proc. International Conference on Sensor Technologies and Applications (SENSORCOMM 2007), Valencia, Spain, 2007.
14. R. E. Jones. Garbage Collection: Algorithms for Automatic Dynamic Memory Management. Wiley, Chichester, July 1996.
15. E. W. Dijkstra, L. Lamport, A. J. Martin, C. S. Scholten, and E. F. M. Steffens. On-the-fly garbage collection: An exercise in cooperation. Commun. ACM, 21(11), 1978.
16. K. A. Mohammed-Ali. Object-Oriented Storage Management and Garbage Collection in Distributed Processing Systems. PhD thesis, Royal Institute of Technology, Dept. of Computer Systems, Stockholm, Sweden, 1984.
17. B. Zorn. Barrier methods for garbage collection. Technical Report CU-CS-494-90, University of Colorado, Boulder, 1990.
18. A. L. Hosking and R. L. Hudson. Remembered sets can also play cards. In ACM OOPSLA,Äô93 Workshop on Memory Management and Garbage Collection, Washington, DC, 1993.
19. D. Tarditi. Compact garbage collection tables. In Tony Hosking, editor, ISMM 2000 Proceedings of the Second International Symposium on Memory Management, volume 36(1) of ACM SIGPLAN Notices, Minneapolis, MN, October 2000.
20. K. Marquet and G. Grimaud. A DSL approach for object memory management of small devices. PPPJ 2007: Proc. International Conference on Principles and Practices of Programming in Java, Lisboa, Portugal, 2007.
21. G. Salagnac, C. Rippert, and S. Yovine. Semi-automatic region-based memory management for real-time java embedded systems. In RTCSA ,Äô07 : Proceedings of the 13th IEEE International Conference on Embedded and Real-Time Computing Systems and Applications, Washington, DC, USA, 2007.

Chapter 17

Efficient Computation of Min and Max Sensor Values in Multihop Networks

Nuno Pereira[1], Björn Andersson[1], Eduardo Tovar[1] and Paulo Carvalho[2]
[1]*IPP-HURRAY Research Group, CISTER/ISEP, Polytechnic Institute of Porto, Porto, Portugal*
[2]*Department of Informatics, University of Minho, Braga, Portugal, bandersson@dei.isep.ipp.pt*

Abstract Consider a wireless sensor network (WSN) where a broadcast from a sensor node does not reach all sensor nodes in the network; such networks are often called multihop networks. Sensor nodes take individual sensor readings, however, in many cases, it is relevant to compute aggregated quantities of these readings. In fact, the minimum and maximum of all sensor readings at an instant are often interesting because they indicate abnormal behavior, for example if the maximum temperature is very high then it may be that a fire has broken out. In this context, we propose an algorithm for computing the min or max of sensor readings in a multihop network. This algorithm has the particularly interesting property of having a time complexity that does not depend on the number of sensor nodes; only the network diameter and the range of the value domain of sensor readings matter.

Keywords Transducers, Data processing, Large-scale sensor networks, MAC protocol

17.1 Introduction

Wireless sensor networks (WSN) often take many sensor readings of the same type (for example, temperature readings), and instead of knowing each individual reading it is more significant to know aggregated quantities of these sensor readings. For example, each sensor node senses the temperature at its location, and the goal is to know the maximum temperature among all nodes at a given moment.

Several solutions for data aggregation have been proposed for multihop networks. Typically, nodes self-organize into a convergecast tree with a base station at the root [1,2]. Leaf nodes broadcast their data. All other nodes wait

N. Martínez Madrid, R.E.D. Seepold (eds.), *Intelligent Technical Systems*, Lecture Notes in Electrical Engineering 38, DOI 978-1-4020-9823-9_17,

until they have received a broadcast from all of their children; a node aggregates the data from its children and makes a single broadcast. Techniques have been proposed for computing useful aggregated quantities such as minimum and maximum values, the number of nodes and the median among a set of sensor nodes. They offer good performance because they exploit the opportunities for parallel transmission, and the processing enroute makes the transmitted packet typically smaller than the sum of the size of the incoming packets.

Despite these optimizations, the performance is still inhibited by the fact that in a single broadcast domain, at most one packet can be sent and hence the time-complexity still depends on the number of sensor nodes. This is particularly problematic for dense networks, where even a small broadcast domain (covering an area $<10\ m^2$) may contain several tens to a few hundred sensor nodes. In order to improve performance to another level, it is necessary to design distributed algorithms that circumvent this limitation.

In this paper, we propose an algorithm for computing the min or max of sensor readings in a multihop network. This algorithm has the particularly interesting property of having a time complexity that does not depend on the number of sensor nodes; only the network diameter and the range of the value domain of sensor readings matter.

We consider this result to be significant because: (i) a significant number of sensor networks are designed for large scale, dense networks and it is exactly for such scenarios that our algorithms excel and (ii) the techniques that we use depend on the availability of a prioritized MAC protocol that supports a very large range of priority levels and is collision-free assuming that priorities are unique, and such a protocol has recently been proposed, implemented and tested on a sensor network platform [3].

The remainder of this paper is structured as follows. Section 17.2 starts by providing an introduction on wireless bit dominance and an application background introducing the main idea of how a prioritized MAC protocol can be used, focusing on a single broadcast domain. The final subsection of Section 17.2 discusses some related work. Section 17.3 presents the new algorithm which offers a time-complexity that is independent of the number of sensor nodes. Finally, Section 17.4 draws the conclusions.

17.2 Preliminaries and Motivation

17.2.1 The Wireless Bit Dominance Approach

The basic premise for this work is the use of a prioritized MAC protocol for wireless medium. This implies that the MAC protocol assures that of all nodes contending for the medium at a given moment, the ones with the highest priority gain access to it. As a result of the contention for the medium, all participating nodes will have knowledge of the winner's priority. This is inspired by Dominance/Binary-Countdown protocols [4], implemented for wired networks in the widely used CAN bus [5].

In our prioritized MAC protocol for wireless medium, lower values mean higher priority, which is also similar to Dominance/Binary-Countdown protocols. However, such protocols assume that priorities are unique. We do not make that assumption.

The prioritized MAC protocol (inspired by Dominance/Binary-Countdown protocols) proposed [3] exhibits this behavior for wireless channels. In this protocol, the nodes start by agreeing on an instant when the contention resolution phase, named tournament, starts. Then nodes transmit the priority bits starting with the most significant bit. A bit is assigned a time interval. A node contends with a dominant bit ("0"), then a carrier wave is transmitted in this time interval; if the node contends with a recessive bit ("1"), it does not transmit but listens. At the beginning of the tournament, all nodes have the potential to win, but if a node contends with a recessive bit and perceives a dominant bit then it withdraws from the tournament and cannot win. If a node has lost the tournament then it continues to listen in order to know the priority of the winner. When a node finishes sending all priority bits without hearing a dominant bit when it transmitted a recessive bit, then it has won the tournament and clearly knows the priority of the winner. Hence, lower numbers represent higher priorities. The work developed [3] includes the precise definition of the protocol timing parameters and accounts for real-world non-idealities such as clock inaccuracies, time of flight, time for detection of carrier pulses or processing delays, and also presents an implementation of the protocol in real-world platforms. Although the proof-off-concept implementation of this prioritized MAC protocol for wireless networks introduces a significant amount of overhead, this overhead is, to a large extent, due to the transition time between transmission and reception, which is essentially a technological parameter, as witnessed by the fact that the Hiperlan standard [6] required a switching time of 2 μs.

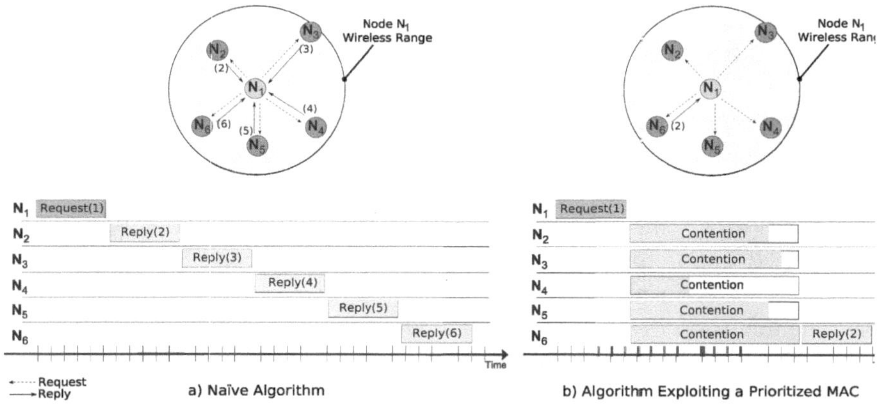

Fig. 17.1 Computing min and max in a single broadcast domain

17.2.2 Motivating Scenario

The focus of this paper will be on exploiting a prioritized MAC protocol as described in the previous subsection. We show that the availability of such a protocol enables efficient distributed computations of aggregated quantities in WSN.

The problem of computing aggregated quantities in a single broadcast domain can be solved with a naïve algorithm: every node broadcasts its sensor reading. Hence all nodes know all sensor readings and then they can compute the aggregated quantity. This has the drawback that in a broadcast domain with m nodes, at least m broadcasts are required. We address the case of WSN designed for large scale, dense networks [7,8]. Under such premise, the naïve approach is inefficient, causing a large delay and energy waste.

Let us consider the simple application scenario depicted in Fig. 17.1(a), where a node (node N_1) needs to know the minimum temperature reading among its neighbors. Let us assume that no other node attempts to access the medium before this node. A naïve approach would imply that N_1 broadcasts a request to all its neighbors and then waits for the corresponding replies from them. As a simplification, assume that nodes have set up a scheme to orderly access the medium in a time division multiple access (TDMA) fashion, and that the initiator node knows the number of neighbor nodes. Then N_1 can compute a waiting timeout for replies based on this knowledge. Clearly, with this approach, the execution time depends on the number of neighbor nodes (m).

Consider now that a prioritized MAC protocol such as the one described in the beginning of this section is available. This alternative would allow an approach as depicted in Fig. 17.1(b). Assume that the range of the analog to digital converters (ADC) on the sensor nodes is known, and that the MAC protocol can, at least, represent as many priority levels. Now, to compute the minimum temperature among its neighbors, node N_1 needs to perform a broadcast request that will trigger all its neighbors to contend for the medium using the prioritized MAC protocol. If neighbors access the medium using the value of their temperature reading as the priority, the priority winning the contention for the medium will be the minimum temperature reading. (The different lengths of the gray bars inside the boxes depicting the contention in Fig. 17.1(b) represent the amount of time that the node actively participated in the medium contention). With this scheme, more than one node can win the contention for the medium. But considering that as a result of the contention, nodes will know the priority of the winner, no more information needs to be transmitted by the winning node.

In this scenario, the time to compute the minimum only depends on the time to perform the contention for the medium, not on m.

A similar approach can be used to compute the maximum temperature reading. Instead of directly coding the priority with the temperature reading, nodes will use the bitwise negation (change every bit of the temperature reading to its opposite value) of the temperature reading as the priority. Upon completion of the medium access contention, given the winning priority, nodes perform bitwise negation to know the maximum temperature.

17.2.3 Previous Work

A prioritized MAC protocol is useful to schedule real-time traffic [3] and it can support data dissemination when topology is unknown [9]. In this paper we will discuss how to efficiently compute aggregated quantities using a prioritized MAC protocol. Distributed calculations have been performed in previous research. It has been observed [10,11] that nodes often detect an event and then need to spread the knowledge of this event to their neighbors [10]. This is called one-to-k communication [10] because only k neighbors need to receive the message. After that, the neighbor nodes perform local computations and report back to the node that made the request for 1-to-k communication. This reporting back is called k-to-1 communication. Algorithms for both 1-to-k and k-to-1 communication are shown to be faster than a naïve algorithm but, unfortunately, the time-complexity increases as k increases. On a single broadcast domain, our algorithms compute a function f and take parameters from different nodes, making the result available to all nodes. In this respect, it is similar to the average calculations in [12]. However, our algorithms are different from others [10–12] as our algorithms have a time-complexity independent of the number of nodes.

One way to use these algorithms is to encapsulate them in a query processor for database queries. Query processors for sensor networks have been studied in previous work [2,13] but they are different in that they do not compute aggregated quantities as efficiently as our approach. They assume one single sink node and that the other nodes should report an aggregate quantity to this sink node. The sink node floods its interest in the data it wants into the network and this also causes nodes to discover the topology. When a node has new data, it broadcasts this data; other nodes hear it, then it is routed and combined so that the sink node receives the aggregated. These works exploit the broadcast characteristics of the wireless medium but they do not make any assumption on the MAC protocol (and hence they do not take advantage of the MAC protocol). One important aspect of these protocols is to create a spanning tree. It is known that computing an optimal spanning tree for the case when only a subset of nodes can generate data is equivalent to finding a Steiner-tree, a problem known to be NP-hard (the decision problem is NP-complete, see page 208 in [14]). For this reason, approximation algorithms have been proposed [15,16]. However, in the average case, very simple randomized algorithms perform well [17]. Since a node will forward its data to the sink using a path which is not necessarily the shortest path to the sink, these protocols cause an extra delay. Hence, there is a trade-off between delay and energy-efficiency. To make this trade-off, a framework based on feedback was developed for computing aggregated quantities [18]. Techniques to aggregate data in the network such that the user at the base station can detect whether one node gives faked data has been addressed as well [19].

Common to these previous works is that the time-complexity increases with the number of sensor nodes. It is clearly desired to create an algorithm which can compute certain quantities, such as MIN and MAX. We have already seen an idea (in Section 17.2.2) how such computations can be performed. But it is desired to do so also in networks with multiple broadcast domains. We will do so in the next section.

17.3 The New Algorithm

It should be clear that the algorithms for computing *min* and *max* in a single broadcast domain (presented in Section 17.2) do not work in a multihop network. In this section, we will extend them.

We assume that nodes are statically placed in a physical location, and that the communication range (R_{co}) is the maximum range at which two nodes N_i and N_j can communicate reliably and the interference range (R_{it}) is the maximum range between nodes N_j and N_k such that simultaneous transmissions to N_j will collide with N_k. We assume that $R_{it} \leq 2R_{co}$. We also assume that time is slotted such that all nodes know the time when a timeslot begins and they also know the identifier of the timeslot. One way to implement that is to use a sensor node platform that is equipped with an Amplitude Modulation (AM) receiver that detects signals from an atomic clock. Such AM receivers are used in the FireFly sensor platform [20] and it receives time-sync signals with a continental wide coverage. Two of them are located in Europe; one of them [21] is located in USA. It is assumed that the duration of the timeslot is equal to the time it takes to run a tournament in the MAC protocol. In order to simplify the discussion, we focus on the computation of min of sensor readings; the max of sensor readings can be designed analogously.

It is also assumed that all sensor nodes know when the computation should start, and do it periodically (for example, let all nodes start this computation at the beginning of a timeslot such that the identifier of the timeslot is divisible by 100). This is sensible for applications that continuously detect fire. But in a multi-tiered architecture, where some nodes have a longer communication range, it is preferable to allow more high-powered sensor nodes to initiate a computation; this assumes that those high powered sensor nodes have a communication range that covers the entire network.

The algorithm is composed of two main steps. At setup time, a topology discovery algorithm is executed to partition the network such that all nodes in each partition are in the same broadcast domain. Then, during runtime, nodes find the minimum sensor reading in all partitions and communicate these values to the leader.

17.3.1 Setup

The setup procedure must partition the network such that (i) each partition forms a single broadcast domain, (ii) a partition leader for each partition is selected, (iii) the partition leaders form a connected distributed set and (iv) to each partition is given a timeslot ensuring that no interfering partitions are active at the same time.

We start this procedure by selecting the partition leaders. To do this we select a *Minimum Virtual Dominating Set* (MVDS) as introduced in [22]. A *Dominating Set* (DS) is a subset of nodes where each node (of the entire graph) is either in the dominating set or is a neighbor to a node in the dominating set. If the set has the minimum cardinality, then it is said to be a *Minimum Set*. To guarantee that all nodes in a partition are in the same broadcast domain, we use a *virtual* range, and thus we construct a MVDS that is the minimum set of nodes required to perform the data aggregation, observing the restrictions (i) to (iii) above.

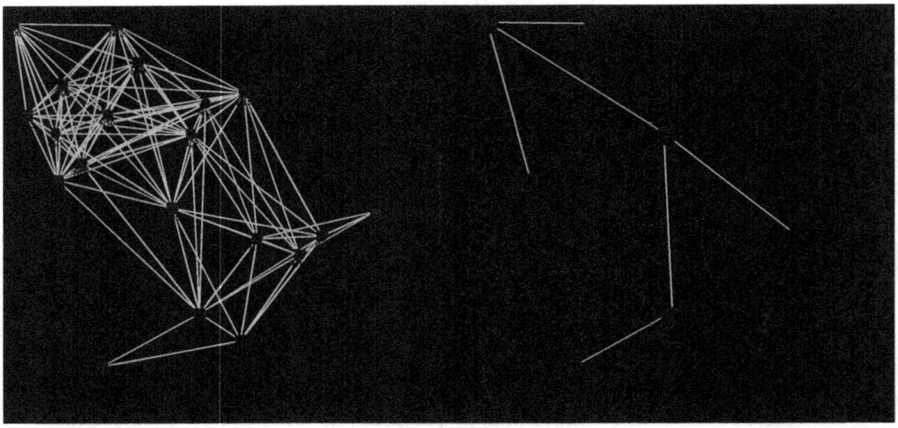

Fig. 17.2 Illustration of the MVDS construction algorithm

The details of the algorithm to construct the MVDS can be found in [22]. It is a distributed algorithm with a *propagation phase* that forms the partitions and colors the nodes according to their functionality (*black* if the node is a partition leader or *red* if it is a slave member of a partition), and a *response phase*, where the topology information is delivered to the leader node. In the beginning of the algorithm, all nodes are *white*. The node starting the algorithm (the leader) colors itself black and broadcasts a message with its color. Nodes within the virtual range of the black node become red and nodes that receive the broadcast but are outside the virtual range become blue.[1] After a time interval that is inversely proportional to the distance from the black node, both red and blue nodes forward the message, if they have not done so. Upon being colored, all blue nodes start a timer to become black. This algorithm approximates the solution for a MVDS(r) composed of the nodes colored black, where r is the virtual range used. It is important to note that, in this work, we select r as a function of the

[1] We assume that distances can be approximated; this can be done, for example, using the signal strength in the received packets.

communication range such that all nodes in each partition are in the same broadcast domain. Based on our assumptions about the communication range, we can define $r = R_{co}/2$.

A possible selection made by the algorithm is illustrated in Fig. 17.2. Figure 17.2(a) presents the positions and connectivity of the network. The different partitions formed are also depicted in Fig. 17.2(a) by representing the nodes in the same partition similarly. Figure 17.2(b) depicts the partition leaders selected by the algorithm and their respective virtual ranges.

After running the propagation phase of the MVDS construction algorithm, the nodes selected as partition leaders report back to the leader the information about the topology of the network. This topology information is used by the leader to assign a timeslot to each partition such that the timeslot is unique from any 1 or 2-hop neighbors.

Algorithm 1 Computing MIN

1. Each sensor nodes N_i takes a sensor reading. Let v_i denote this sensor reading.
2. Each node N_i in $PART_j$ waits until the time slot SLOT($PART_j$) and then it sends an empty packet with the priority given by v_i. After the tournament, the partition leader knows the minimum v_i. Let $winnerprio_i$ denote this value.
3. Communicate the results $winnerprio_i$ from partition leaders to the leader.
4. The leader takes the min of all $winnerprio_i$ that it receives. This minimum is the minimum of all sensor readings.

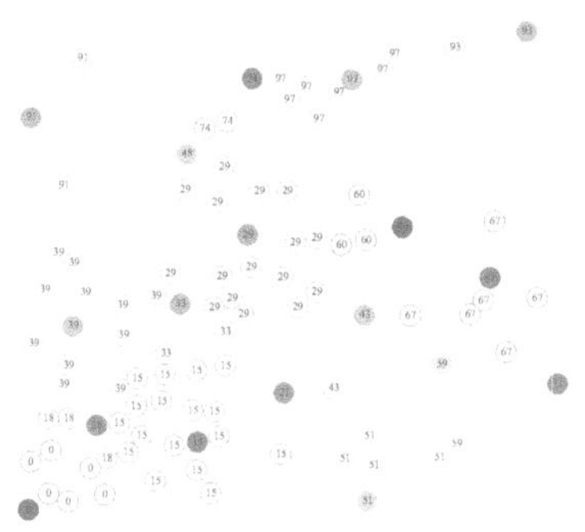

Fig. 17.3 Partitioning and Partition Leaders for an Example Network

17.3.2 Runtime

At runtime, nodes have to find the minimum value within each partition, and then the partition leaders deliver these minimum values to the leader.

Algorithm 1 provides the sequence of steps the nodes take during runtime.

While the minimum values are routed to the leader, partition leaders can perform simple processing and avoid forwarding min or max values that are higher or lower than values previously transmitted.

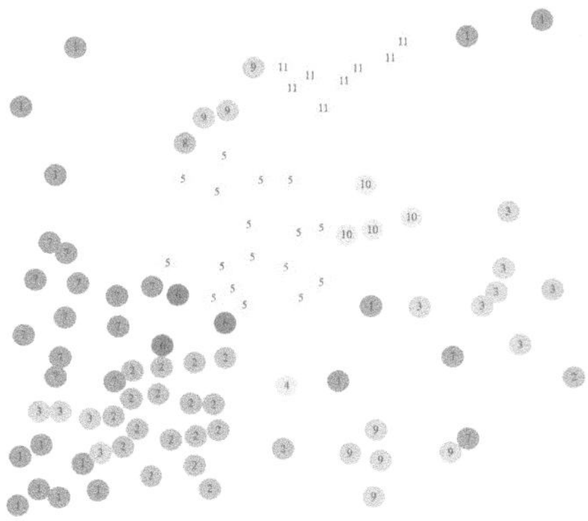

Fig. 17.4 Timeslots assigned to partitions

Fig. 17.5 Result After Timeslot 1

Fig. 17.6 Each sensor node and the original sensor reading

Fig. 17.7 Result after timeslot 11

17.3.3 Running an Example

We will illustrate the algorithm with a simple example. Figure 17.3 shows a sensor network consisting of 100 nodes.

Let us consider the algorithm that is run when the sensor network is deployed (as described in Section 17.3.1). The algorithm partitions the network and selects the corresponding partition leaders. Figure 17.3 depicts the partition leaders with a solid grey circle, the numbers in each node are the partition-ids to which the node belongs (partition-ids are assigned according to the partition leader address).

Then timeslots are assigned to each partition such that if two sensor nodes, in different partitions but in the same timeslot, broadcast simultaneously, then there is no collision. Figure 17.4 shows the timeslot assigned to each node. As illustrated, there are 11 different timeslots.

Let us consider the algorithm that is executed at runtime. Figure 17.5 shows the temperature readings in all nodes. Nodes compete for the channel using their temperature readings as the priority and nodes do this in their assigned timeslot. After this competition, all nodes know the minimum of temperature in the partition. Figure 17.6 shows the result after the first timeslot. Observe that the nodes depicted in solid grey circles have all the same value within the corresponding partitions. This is because these nodes were assigned timeslot 1 and the values depicted are the minimum values in each partition, spread to all sensor nodes in the same partition. After 11 timeslots, all nodes have broadcasted their temperature reading. Figure 17.7 shows the result after the 11:th timeslot. Now, every leader of a partition knows the minimum temperature in the partition. Finally, nodes perform convergecast to the leader of the entire network. Observe that, due to the setup phase, nodes are organized in partitions where member nodes know their partition leaders and partition leaders known the other parent partition leaders who can forward message towards the leader node. Thus performing convergecast is trivial. After the convergecast, the leader knows that the minimum temperature in the entire network is 5.

To further illustrate why the algorithm is fast, a randomly generated network with 1000 nodes is depicted in Figure 17.8. In this figure, the 77 partition leaders are depicted with solid circles, slightly bigger than the other nodes. In this network 17 unique timeslots are needed. By this example, we can observe that our scheme scales well.

So far we have assumed that all transceivers can only transmit in a pre-specified channel. But many wireless standards, such as 802.11, allow a transceiver to transmit on any channel. This feature can be used advantageously by assigning each partition its own channel (instead of assigning a timeslot to a partition) and this reduces the time required to perform step 2 in Algorithm 1.

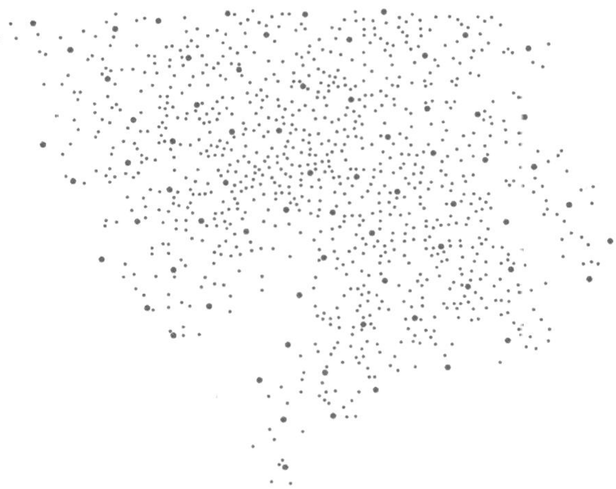

Fig. 17.8 Large-scale network example

17.4 Conclusions

We have shown how to use and take advantage of a prioritized MAC protocol to compute aggregated quantities efficiently. The algorithms designed to exploit such MAC protocol have a time-complexity that is independent of the number of sensor nodes. This is clearly important for WSN applications that operate under real-time constraints. As the lower time taken to perform computations allows nodes to be awake for shorter periods (longer sleeping times), and thus energy consumption is also reduced, providing nodes a longer life-time.

For future research, some questions remain open: (i) Can other methods for partitioning the network make this technique perform better? (ii) Can a similar technique be used to compute more complex aggregated quantities (such as COUNT, MEDIAN and interpolation)? (iii) Is the technique sufficiently reliable for large-scale systems?

Acknowledgement This work was partially funded by CONET, the Cooperating Objects Network of Excellence, funded by the European Commission under FP7 with contract number FP7-2007-2-224053, the ARTIST2 Network of Excellence on Embedded Systems Design and by the Portuguese Science and Technology Foundation (Fundação para Ciência e Tecnologia - FCT).

References

1. J. Gehrke and S. Madden, "Query Processing for Sensor Networks," IEEE Pervasive Computing, Vol. 3, pp. 46–55, January–March 2004.
2. S. Madden, M. J. Franklin, J. M. Hellerstein, and W. Hong, "TAG: a Tiny AGgregation service for ad-hoc sensor networks," in 5th symposium on Operating systems design and implementation (OSDI '02), 2002, pp. 131–146
3. N. Pereira, B. Andersson, and E. Tovar, "WiDom: A Dominance Protocol for Wireless Medium Access," IEEE Transactions on Industrial Informatics, Vol. 3, May 2007.
4. A. K. Mok and S. Ward, "Distributed Broadcast Channel Access," Computer Networks, Vol. 3, pp. 327–335, 1979.
5. Bosch, "CAN Specification, ver. 2.0, Robert Bosch GmbH, Stuttgart," 1991.
6. ETSI, " TS 101 475 V1.3.1:," Broadband Radio Access Networks (BRAN);HIPERLAN Type 2; Physical (PHY) layer.
7. A. Arora, "ExScal: Elements of an Extreme Scale Wireless Sensor Network," in Proceedings of the 11th IEEE International Conference on Embedded and Real-Time Computing Systems and Applications (RTCSA'05), Washington, DC, USA, 2005, pp. 102–108.
8. W. R. Heinzelman, A. Chandrakasan, and H. Balakrishnan, "Energy-Efficient Communication Protocol for Wireless Microsensor Networks," in Proceedings of the 33rd Hawaii International Conference on System Sciences (HICSS'00), Maui, U.S.A., 2000, pp. 3005–3014.
9. B. Andersson, N. Pereira, and E. Tovar, "Disseminating Data Using Broadcast when Topology is Unknown," in 26th IEEE Real-Time Systems Symposium (RTSS'05), Work-in-Progress Session, 2005, pp. 61–64.
10. R. Zheng and L. Sha, "MAC Layer Support for Group Communication in Wireless Sensor Networks," Department of Computer Science, University of Houston UH-CS-05-14, July 21 2005.
11. K. Jamieson, H. Balakrishnan, and Y. C. Tay, "Sift: a MAC Protocol for Event-Driven Wireless Sensor Networks," in Third European Workshop on Wireless Sensor Networks (EWSN), Zurich, Switzerland, 2006.
12. D. S. Scherber and H. C. Papadopoulos, "Distributed computation of averages over ad hoc networks," IEEE Journal on Selected Areas in Communications, Vol. 23, pp. 776–787, April 2005.
13. Y. Yao and J. Gehrke, "Query processing in sensor networks," in Proceedings of the 1st Biennial Conference on Innovative Data Systems Research (CIDR'03), 2003.
14. W. Jianping, M. McDonald, M. Brackstone, L. Yangying, and G. Jingjun, "Vehicle to vehicle communication based convoy driving and potential applications of GPS," in

Proceedings of the 2nd International Workshop on Autonomous Decentralized Systems, 2002, pp. 212–217.

15. B. Krishnamachari, D. Estrin, and S. B. Wicker, "The Impact of Data Aggregation in Wireless Sensor Networks," in Proceedings of the 22nd International Conference on Distributed Computing Systems (ICDCS'02), 2002, pp. 575–578.
16. C. Intanagonwiwat, D. Estrin, R. Govindan, and J. Heidemann, in Proceedings of the 22nd International Conference on Distributed Computing Systems (ICDCS'02), Washington, DC, USA, 2002, p. 457.
17. E. M., G. A., G. R., and M. R., "Scale-free Aggregation in Sensor Networks," Theoretical Computer Science, Vol. 344, pp. 15–29, 2005.
18. T. Abdelzaher, T. He, and J. A. Stankovic, "Feedback Control of Data Aggregation in Sensor Networks," in Proceedings of the 43rd IEEE Conference on Decision and Control (CDC'04), 2004, Vol.2, pp. 1490–1495.
19. B. Przydatek, D. Song, and A. Perrig, "{SIA}: Secure information aggregation in sensor networks," in Proceedings of the 1st ACM International Conference on Embedded Networked Sensor Systems (SenSys'03), 2003, pp. 255–265.
20. R. Mangharam, A. Rowe, and R. Rajkumar, "FireFly: A Cross-Layer Platform for Wireless Sensor Networks," Real Time Systems Journal, Special Issue on Real-Time Wireless Sensor Networks, Vol. 37, pp. 183–231, 2007.
21. NIST Radio Station WWVB," http://tf.nist.gov/stations/wwvb.htm.
22. B. Deb, S. Bhatnagar, and B. Nath, "Multi-resolution state retrieval in sensor networks," in Sensor Network Protocols and Applications, 2003. Proceedings of the First IEEE. 2003 IEEE International Workshop on, 2003, pp. 19–29.

Chapter 18

A Low-Cost FPGA-Based Embedded Fingerprint Verification and Matching System

Maitane Barrenechea, Jon Altuna, Mikel Mendicute and Javier Del Ser
University of Mondragon, Loramendi 4, 20500 Arrasate-Mondragon, Spain
mbarrenetxea@eps.mondragon.edu, jaltuna@eps.mondragon.edu, mmendikute@eps.mondragon.edu, jdelser@eps.mondragon.edu

Abstract The development of a fingerprint verification system on a low-cost embedded platform still remains an open issue in nowadays biometrics. In order to shed light on this field, the contribution shown in this manuscript describes a low-cost fingerprint minutiae extraction and matching system based on a Spartan3 field-programmable gate array (FPGA) with an embedded Leon2 open core processor. The proposed system architecture incorporates a floating point unit (FPU) and a discrete Fourier transform (DFT) coprocessor which accelerates the minutiae extraction process. The verification algorithm is based on the NFIS (NIST Fingerprint Image Software) version 2 open source software developed by the National Institute of Standards and Technology (NIST). A software enhancement algorithm has also been included to further accelerate the minutiae extraction process. The results on execution time reduction and FPGA occupation for different system configurations show that the proposed architecture improves substantially the performance of the baseline system

Keywords Fingerprint, Biometric, FPGA, Embedded processor, Leon2.

18.1 Introduction

In nowadays society identity verification is becoming a crucial issue in several business sectors such as access or border control. Due to this fact, a new field known as biometrics has emerged, which uses some unique physiological or behavioral characteristics, not shared by any other individual, to positively identify a person. Examples of physical characteristics include fingerprints, hand measurements, facial patterns, eye retinas and irises, whereas examples of mostly behavioral characteristics

N. Martínez Madrid, R.E.D. Seepold (eds.), *Intelligent Technical Systems*, Lecture Notes in Electrical Engineering 38, DOI 978-1-4020-9823-9_18,

include signature, gait and typing patterns. In this context, fingerprint based verification is one of the most used biometric systems due to its easiness of acquisition, high distinctiveness, persistence and acceptance by the public [1].

A comparison of different biometric systems is show in Table 18.1. The comparison is made by evaluating seven different factors [2]: (a) universality, (b) uniqueness, (c) permanence, (d) collectability, (e) performance, (f) acceptability and (g) circumvention.

Table 18.1 Comparison of several biometric systems (H=high, M=medium, L=low)

	a	*b*	*c*	*d*	*e*	*f*	*g*
Fingerprint	M	H	H	M	H	M	H
Retina	H	H	M	L	H	L	H
Iris	H	H	H	M	H	L	H
Face	H	L	M	H	L	H	L
Facial thermogram	H	H	L	H	M	H	H
Hand Geometry	M	M	M	H	M	M	M
Hand Vein	M	M	M	M	M	M	H
Signature	L	L	L	H	L	H	L
Keystroke dynamics	L	L	L	M	L	M	M

Furthermore, fingerprint biometric systems can be split into two different approaches: pattern classification and minutiae detection. The former associates a type of ridge structure, such as loop, tented arch or whorl, and its particular features to each fingerprint, whereas the latter approach extracts the minutiae from a given fingerprint image. Minutiae are points of interest in a fingerprint, such as ridge endings or bifurcations. For the work described in this article the minutiae detection approach has been selected due to its superior resistance to physical degradations and the high speed of the matching algorithms associated to this kind of fingerprint-based verification system.

This document describes the design flow and implementation results of a low-cost embedded system for fingerprint verification. The proposed system consists of a 32-bit SPARC Leon2 processor, a fingerprint image sensor, a signal processing hardware accelerator and an FPU. It is worth remarking that the minutiae extraction module is open to work with any fingerprint sensor, so a change in the image capture driver is enough for any new fingerprint acquisition device.

Similar FPGA-based fingerprint verification systems have been developed and proposed in the literature [3–5]. From the system architecture point of view, the Thumbpod project [3] is probably the most important reference due to its similarities to the platform presented in this paper. Both systems have been built upon the Leon2 soft-processor, although for our system implementation we have selected a low-cost Spartan3 FPGA as the core of the design, while a more expensive Virtex II device was chosen in the Thumbpod project. Moreover, a hardware FPU coprocessor enables our system to accurately perform high speed floating point operations in contrast with the fixed-point refinement required in the Thumbpod project. Regarding software development issues, both projects are based on the NFIS open source software. Nevertheless, the verification system proposed in this paper has its roots in the enhanced version 2 (NFIS2) of this algorithm and uses the specified input fingerprint image format (500 pixels per inch (ppi) and 256 greyscale images) for its optimum performance. On the other hand, the Thumbpod project algorithm employs low quality images (3 bits per pixel) as an input pattern to execute the NFIS version 1 minutiae extraction flow. The proposed system is also open to most fingerprint sensors in contrast to other platforms which have been customized for a specific fingerprint sensor. As for the matching algorithm is concerned, the BOZORTH3 algorithm has been implemented [6].

This article is organized as follows: In Section 18.2, the software architecture for the minutiae extraction and matching algorithm targeted to a Leon2 based platform is described. Section 18.3 details the proposed HW architecture of the design, emphasizing the floating point operation acceleration achieved by means of an FPU and a DFT co-processing engine. Section 18.4 deals with the proposed optimization approaches. Section 18.5 shows the main timing and complexity results for different system configurations and finally, Section 18.6 provides some concluding remarks.

18.2 Software Architecture

The fingerprint authentication algorithms developed for the target system are based on some routines of the NFIS2 collection. Specifically, custom versions of the MINDTCT and BOZORTH3 packages have been developed for the minutiae acquisition and matching, respectively.

The algorithm and the implemented software parameters have been designed and set for an optimum performance with 256 grayscale and 500 ppi scanned images. These input image features match perfectly with those provided by the most relevant fingerprint sensors, such as the Fujitsu MBF200, which has been chosen for this research work.

18.2.1 Software Implementation on a Leon2 Platform

The original software was designed and tested to be run on a Linux operating system and compiled with gcc. A bare-C cross-compiler and the GRMON debug monitor from Gaisler Research have been used in this project. Dynamically allocated data arrays have been used to store intermediate results depending on the application requirements.

On the other hand, although the original software only accepts input image files in ANSI/NIST, WSQ, JPEGB, JPEGL and IHEAD formats, the selected fingerprint sensors provide images in RAW format, for which the algorithm has been modified and adapted so that only this type of image files are accepted.

18.2.2 Minutiae Extraction Algorithm

The MINDTCT software has been designed in a modular fashion, and as a result, each step in the algorithm is mainly executed in a subroutine. Only those strictly necessary modules for XYT (position, direction and quality) formatted output minutiae list generation have been selected from the original algorithm. The functional steps executed in the minutiae extraction algorithm are the following:

- Input: Fingerprint RAW image.
  - **a.** Generation of image maps.
  - **b.** Binarization.
  - **c.** Minutiae detection.
  - **d.** False minutiae removal.
  - **e.** Minutiae quality assessment.
- Output: Minutiae in XYT format.

Degraded fingerprint areas, which are prone to give as a result erroneous minutiae, are identified in the image map generation phase. To this end, unreliable image zones are detected by means of three image maps which are depicted in Fig. 18.1:

- Low contrast map (LCM): Marks low contrast areas which mainly correspond with the background of the image or smudges in the fingerprint.
- Low ridge flow map (LFM): Identifies those image areas where the dominant ridge flow could not be determined initially.

- High curvature map (HCM): Flags high curvature areas in the image, such as the fingerprint core or possible delta regions.

Fig. 18.1 Generated image maps by the MINDTCT algorithm from left to right: low contrast map, low ridge flow map and high curvature map

As a combination of these three features a quality map is derived, which is depicted in Fig. 18.2. This map assigns one of five possible quality levels, ranging from 0 to 4 in increasing order of quality, to each of the blocks in the image

In this phase of the algorithm one of the fundamental maps for the minutiae extraction process is also derived: the directional ridge flow map. For the acquisition of this map, the original image is divided into 8×8 size pixel blocks. A window of 24×24 pixels is defined for each of the image blocks, conformed by the block itself and other surrounding pixels. The window is rotated incrementally in the 16 orientations defined in the algorithm (each of them 11.25 degrees apart) and a DFT is executed at each position. In every orientation, the pixels along each rotated row of the window are summed up to form 16 vectors of row sums. Each one of these vectors is then convolved with eight waveforms of different frequency (two waveforms, sine and cosine, per each of the four ridge widths). The spatial frequency of each waveform discretely represents the width of different ridges and valleys, in such a way that widths of 12, 6, 3 and 1.5 pixels are covered in the algorithm. To determine the dominant ridge flow within a block, the resonance coefficient obtained from the convolution is evaluated. The result for this module of the algorithm is shown in Fig. 18.3.

Once the image maps have been acquired, it is necessary to binarize the fingerprint image so that the minutiae can be extracted. To carry out this process, the previously computed directional ridge flow map is used to determine the binary value assigned to each pixel. After the binarization (Fig. 18.3) the minutiae detection module analyses the binarized image looking for candidate minutiae (ridge ending or bifurcation). However, not all the ridge patterns selected after this procedure correspond to true minutiae, therefore, a false minutiae removing process is carried out. Even after this process, false minutiae may potentially remain in the candidate list. To counteract this fact, a reliability measure is assigned to each minutia based on the quality map and other pixel intensity statistics. The resulting minutiae for the template image are shown in Fig. 18.3.

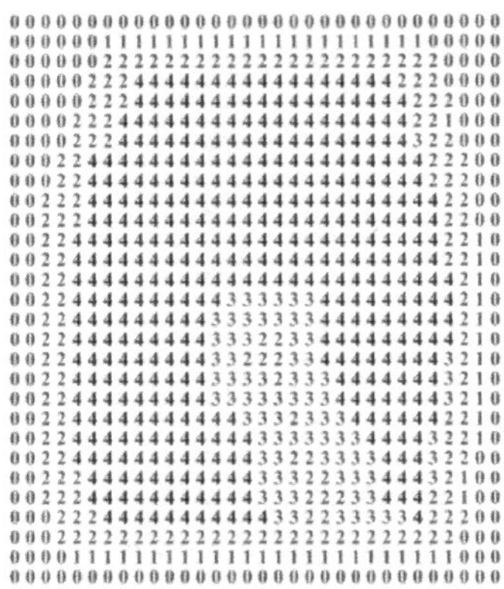

Fig. 18.2 Quality map generated by the MINDTCT algorithm representing the reliability of the areas in the image

Fig. 18.3 Resulting images of the minutiae extraction process from left to right: Ridge direction map, binarized image and the extracted minutiae

18.2.3 Matching Algorithm

The BOZORTH3 matching algorithm, included in the second distribution of NFIS, computes a match score that reflects the similarity degree between a fingerprint minutiae and a template minutiae set, both of them in XYT format. One of the most remarkable features of this algorithm is its invariance to both rotation and translation.

The first step in the algorithm is to construct a comparison table for each one of the input minutiae sets. Relative measures between a minutia and the rest of the minutiae in the same fingerprint are computed and stored in a comparison table. This is what provides the algorithm translation and rotation invariance.

The next step is to look for compatible entries between the two tables. The results of this analysis are stored in a new compatibility table which consists of a list of associations between two possible corresponding minutiae. Each one of these associations represents single links in the compatibility graph.

In the final phase of the matching software flow, the compatibility graph is traversed and clusters are created by linking table entries. Once the traversals are complete, compatible clusters are combined and a match score is computed by accumulation of the linked table entries across the combined clusters.

Generally, a match score greater than forty indicates that both fingerprint and template minutiae belong to the same finger, and so, to the same individual.

18.3 Hardware Architecture

18.3.1 Initial System Architecture

The initial hardware architecture is composed of a 50 MHz fixed-point Leon2 soft-processor with 8 KB of cache memory for data and instructions, all embedded in a GRXC3S1500 board. This processor has been chosen for this application not only because of its high performance and usability, but also due to the fact that it can be obtained under Lesser General Public license (LGPL). According to a report on synthesizable CPU cores [7], where Leon2, MicroBlaze and OpenRISC 1200 where tested under three different hardware configurations and three different benchmarks, Leon2 yielded the best performance per clock cycle for all the scenarios. Moreover, according to the aforementioned report, Leon2 is the processor with the highest usability among the tested CPU cores. This may be due to the VHDL code availability and the TCL/Tk based configuration tool, which facilitates the design of a custom Leon2 based system. The fingerprint image acquisition is performed using a custom-made intellectual property (IP) module connected to the MBF200 fingerprint sensor. This module is attached to the AMBA peripheral bus (APB) and provides the processor with the input fingerprint image. The operation of the sensor is controlled by means of three on-chip registers (control, data and status) generated in the address range allocated for the APB bridge.

18.3.1.1 Running the Application on the Initial System

The original minutiae extraction and matching algorithm was implemented using floating point arithmetic whereas the Leon2 soft-processor is fixed-point. In order to run the program on the target platform the floating-point emulation must be set in the compiler options. This option forces all floating-point operations to be done in software with integer arithmetic.

The execution of the algorithm is successful as for the matching results is concerned but not in terms of execution time. The required computation time for the minutiae extraction is 157 s while the matching process, which is evaluated against a 55 template set, for a one-to-one comparison is carried out in 5–51 s. Even when the results for the extracted minutiae and match score are correct, the program execution delay is unacceptable for a biometric verification system. The excessive execution time is mainly due to the MINDTCT algorithm, and thus, the analysis of the reduction of the time required for minutiae extraction becomes one of the main objectives of this article.

A great amount of floating-point data is used in the MINDTCT module. The emulation of this data format introduces a prohibitive delay in the execution of the program. Hence, an FPU is required to accelerate this process.

The Leon2 processor provides an interface to different FPU blocks, including the Gaisler Reseach FPU (GRFPU), the Meiko FPU by Sun Microsystems, as well as the incomplete LTH FPU [8]. GRFPU has been chosen for the acceleration of the floating point processes because of its high-performance and compliance with the IEEE-754 standard.

The insertion of an FPU in the embedded system leads to a considerable increase in the amount of logic inside the FPGA. This is the reason why a reduction in the processor clock and/or a cutback in the cache memory amount is required. The performance of the biometric verification system has been analyzed for three different system configurations: 31 MHz and 8KB cache memory, 37 MHz and 8KB cache memory and 40 MHz and 4KB cache memory. The rationale for the selection of these three system settings was to test the performance of the system under different clock frequencies and cache memory sizes. The FPGA utilization is almost complete for the Spartan-3 1500 and very similar for the three analyzed hardware configurations, being the 99% of the slices and 53% of the 18 × 18 multipliers occupied for all the tested settings. A 56% of the RAMB16 blocks are allocated for all the system configurations except for the 37 MHz and 8 KB cache memory setting, in which case the occupation is 68%. The results of the aforementioned study are summarized and extended in Section 18.5.

18.3.2 Introducing the GRFPU in the Design

Tests on IEEE-754 compliance drew positive results for the Gaisler Research floating point core, and therefore this module was included in the hardware design. Results for computation time have improved substantially after the FPU insertion, mainly in the execution time required for the MIND-TCT process completion. A 94.14% time reduction has been achieved for the case of 40 MHz and 4 KB cache memory configuration. The execution time for the matching algorithm had a slight improvement.

The timing results for the different system configurations are shown in Table 18.3.

18.4 Performance Optimization

Even if the computation time for the minutiae extraction algorithm has been reduced in a 94.14%, the program completion delay is yet excessive for its implementation in a real commercial system. Therefore, hardware and software optimization approaches are suggested in this section to enhance the efficiency of the minutiae extraction process.

Several timing analyses show that the 75% of the computation time is occupied by the low contrast map, direction map and low flow map generation process. The 92% of the time dedicated to image map computation is needed to generate the directional ridge flow map, as shown in Fig. 18.4 In light of these results, the acceleration of the direction map computation has been the first target in the optimization process.

18.4.1 Hardware Optimization

The process of extracting the direction map begins by dividing the image into 24 × 24 pixel blocks. Every image block is then rotated in the 16 directions defined in the algorithm. The pixels of each row are summed up at each orientation generating 16 row sums per block, which are later convolved with four waveforms of increasing frequency, each of which represents a different ridge width. The result of these computations is a 4 × 16 matrix comprised of the resonance coefficients for the four ridge widths and the 16 directions defined in the algorithm. The direction that generates a higher resonance coefficient is the one to be selected for that image block.

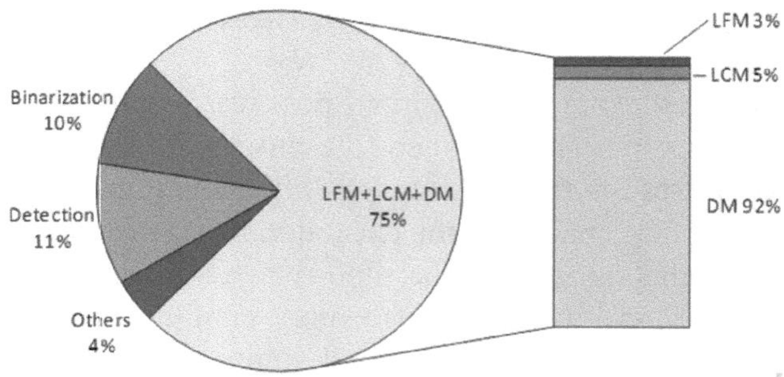

Fig. 18.4 Profiling of the execution time for the different modules of the MINDTCT algorithm

Thorough timing analyses show that the process of multiplying the row sums of the image with the waveforms during the computation of the DFT is the most time-consuming task. All the calculations implied in this process are executed in a sequential fashion which entails a high delay in the acquisition of results from the algorithm. By exploiting the parallelism and pipelining of resources in the FPGA this process can be accelerated to a great extent.

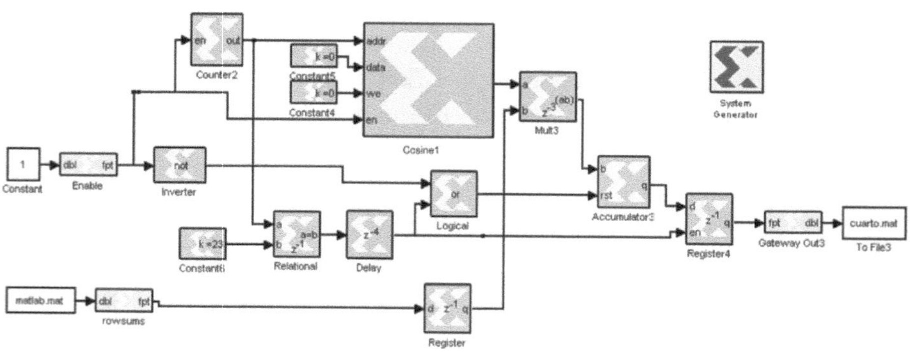

Fig. 18.5 System generator model for one of the eight branches of the hardware accelerator

Taking this into account, a new IP module has been designed using Xilinx System Generator for DSP to perform the multiply and accumulation (MAC) operations in parallel. The input and output ports of the hardware acceleration IP have been designed to interface the APB bus. Figure 18.5 depicts the System Generator model of one of the eight parallel branches of the proposed hardware acceleration IP.

The performance of the new module has been assessed by means of a testbed developed in Matlab/Simulink using System Generator for DSP. The results show that the computation of the direction map of the image could be carried out in 1.29 s for the sample fingerprint image. Once this result is known, the calculation of the time required for the minutiae extraction process is straightforward, showing its reduction from 9.2 s to 4.142 seconds (Table 18.3, configuration E).

The proposed system architecture is depicted in Fig. 18.5.

18.4.2 Software Optimization

As stated in Section 18.2.2, the original algorithm designed by NIST performs the calculation of the ridge direction flow for each block in the image in an independent fashion. As a consequence, it is necessary to perform 16 DFTs in each of the image blocks, which leads to a high number of DFT computations. For example, 16384 DFT calculations should be carried out for a 256×256 image according to the original algorithm.

However, the ridge structure in a fingerprint has a continuous nature, and therefore, adjacent image blocks tend to have similar ridge directions. A software optimization method for the computation of the direction map has been designed based on the approach defined in the Thumbpod proyect [9].

The coefficients for all the 16 directions are computed for the first block in the image. When computing the next block in the row, the directions A-1, A and A+1 will be calculated first, being A the result of the ridge flow direction in the previous image block. If the DFT coefficient for direction A is greater than the other values, in other words a peak value is detected, and a control threshold is exceeded, then a solution has been found. However, if the maximum is found in either of the directions A–1 or A+1, then the DFT value for the adjacent direction is calculated. For instance, if the maximum value was found in the direction A–1, the next direction to be computed would be A–2. This process is repeated until a peak with a greater value than the threshold is found. If after 16 DFT computations no value exceeds the threshold, then the maximum value is selected as the result for that block.

The value of the control threshold has a great impact on the precision of the results as well as on the degree of computational reduction. The effect of several threshold values on the derivation of the direction map is depicted on Fig. 18.6.

For high threshold values, the direction maps derived are very accurate but the acceleration of the process is scarce, and vice versa. The degree of computational reduction, that stems from calculation only a partial subset of all the possible DFT computations, and the loss in accuracy, which refers to the amount of wrong direction blocks in the image, for several threshold values are shown in Table 1.2.

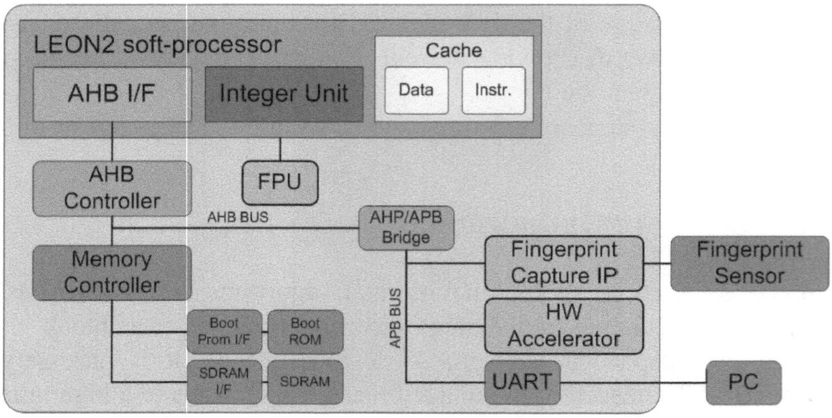

Fig. 18.6 Proposed system architecture based on a Leon2 soft-processor core

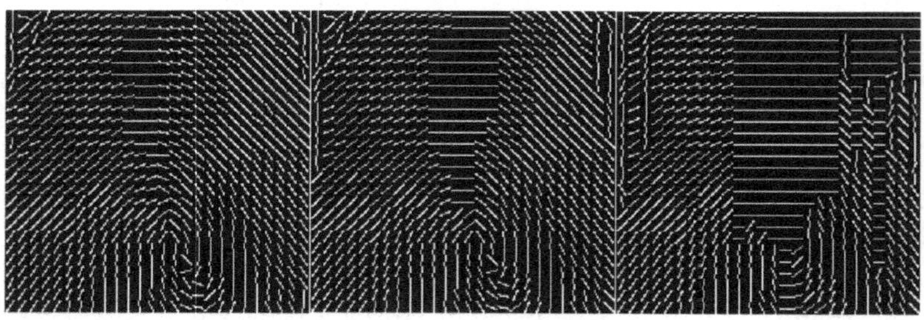

Fig. 18.7 Effect of the value of threshold on the computation and accuracy of the ridge direction map. From left to right: 500×10^6, 60×10^6 and 10×10^6 threshold

The value of the threshold has been set to 60×10^6, due to the good accuracy of the results and the high computational reduction implied. Applying the software acceleration routines to the system, the minutiae extraction process can be further reduced to 3.36 s.

Table 18.2 Effect of the value of threshold on the computational reduction and the accuracy of the direction map

Threshold	500×10^6	200×10^6	100×10^6	60×10^6	10×10^6	5×10^6
Computational reduction	0.0015%	59,69%	68.26%	72.55%	76.37%	78.42%
Loss in accuracy	0%	3.02%	7.41%	11.11%	37.31%	82.17%

18.5 Results

Table 18.2 shows the timing results for several system configurations and different degrees of optimization: (a) 50 MHz clock frequency with 8KB cache memory, (b) 31 MHz clock frequency with 8KB cache memory and FPU, (c) 37 MHz clock frequency with 8KB cache memory and FPU, (d) 40 MHz clock frequency with 4KB cache memory and FPU, (e) 40 MHz clock frequency with 4KB cache memory, FPU and HW optimization, and (f) 40 MHz clock frequency with 4KB cache memory, FPU, HW optimization and SW optimization. Note that the results for the last two configurations are estimated timings only.

Table 18.3 Execution time results for different Leon2-based system configurations (* in respect to configuration a)

	a	*b*	*c*	*d*	*e*	*f*
Execution time (s)	157	12.7	9.5	9.2	4.14	3.36
Time reduction*	–	91.9%	93.9%	94.1%	97.3%	97.8%

18.6 Conclusions

This paper describes the implementation of a fingerprint minutiae extraction and matching algorithm running on a Spartan3 FPGA-based system with an embedded Leon2 soft-processor. The original application developed by NIST has been modified and ported to the target platform. Several tests have been carried out to analyze the performance of the software algorithms with different Leon2 and GRFPU configurations. After the insertion of a floating-point unit, the results on execution time of the algorithm have been reduced in a 94.14% for a 40 MHz and 4 KB cache memory configuration. Further hardware and software optimizations have been proposed to reduce the computation time of the minutia extraction process in a 97.85%.

Acknowledgements This work was supported by the BIOSEG PROFIT Project funded by the Spanish Ministry of Science and Technology. The authors would also like to thank Jiri Gaisler and Richard Pender for their support in setting up the system architecture.

References

1. A.K. Jain, Biometric recognition: How do I know who you are?, in Proceedings of the 12th IEEE Signal Processing and Communications Applications Conference, 3–5, (2004).
2. A.K. Jain, A. Ross and S. Prabhakar, An introduction to biometric recognition, IEEE Transactions on Circuits and Systems for Video Technology 14(1), 4–20 (2004).
3. S. Yang, K. Sakiyama and I. Verbauwhede, A compact and efficient fingerprint verification system for secure embedded devices, in Proceedings of the Asilomar Conference on Signals, Systems and Computers, 2058–2062, (2003).
4. A. Lindoso, L. Entrena and J. Izquierdo, FPGA-based acceleration of fingerprint minutiae matching, in Proceedings of the 3rd Southern Conference on Programmable Logic, 81–83, (2007).
5. M. Lopez Garcia and E. F. C. Navarro, FPGA implementation of a ridge extraction fingerprint algorithm based on a MicroBlaze and hardware coprocessor, in Proceedings of the International Conference on Field Programmable Logic and Applications, 1–5, (2006).
6. C.I. Watson, M.D. Garris, E. Tabassi, et al., User's guide to NIST fingerprint image software 2 (NFIS2), National Institute of Standards and Technology (NIST), 2004.
7. D. Mattson and M. Christensson, Evaluation of synthesizable CPU cores, Master‚Äôs Thesis, Chalmers University of technology, Gothenburg, Sweden, 2004.
8. M. Kasprzyk., Floating Point Unit, Digital IC Project 2001, 2002.
9. S. Yang and I. Verbauwhede, A realtime, memory efficient fingerprint verification system, in Proceedings of the IEEE International Conference on Acoustics, Speech and Signal Processing, V-189-V-192 (2004).

Chapter 19

FPGA-Rootkits

Markus Kucera and Michael Vetter
University of Applied Sciences Regensburg, Regensburg, Germany
markus.kucera@informatik.fh-regensburg.de

Abstract This paper describes the security implications of FPGAs to the Trusted Computing Base of Embedded Systems. It gives an overview of different FPGA architectures and discusses the security measures and shortcoming of modern FPGAs. Furthermore, it shows how an attacker can exploit these shortcomings and integrate rootkit-like code inside the FPGA. After a discussion on possible countermeasures, a description on the different ways a rootkit can be deployed into the FPGA is given.

Keywords Security, FPGA, Rootkits, Trusted computing

19.1 Introduction

The use of FPGAs is on the rise. "In Stant" reports in May 2006 that the market for FPGAs will rise from 2005 1.95 Bn $ to 2.75 Bn $ in 2010 [1].

In many of the new applications security is of high importance.

An obvious example is a FPGA based Cryptographic Coprocessor for either symmetric and/or asymmetric encryption [2].

Another, more exotic, example is the usage of FPGAs to analyze IP packets for an Intrusion Detection System [3].

Good reasons exist to use FPGAs for these applications. FPGAs can perform massive parallel computation at a very high granularity.

Furthermore, modern FPGAs can hold more than 1 Million Gates, therefore multiple tasks can be integrated into one chip, simplifying production and board design.

Moreover, FPGAs can be upgraded/repaired after the production and deployment of a product. When a critical error occurs a new configuration is created and shipped to the costumer.

N. Martínez Madrid, R.E.D. Seepold (eds.), *Intelligent Technical Systems*, Lecture Notes in Electrical Engineering 38, DOI 978-1-4020-9823-9_19,

Finally, partial runtime reconfiguration increases the efficiency of the FPGA by changing the device configuration on demand.

However, the growing flexibility of FPGAs raises questions about the security implications for the whole system.

This paper gives an overview over Rootkits in general and discusses security measures and shortcoming of modern FPGAs. Furthermore, it shows how an attacker can exploit these shortcomings to integrate Rootkit-like code in the FPGA. To the authors' knowledge no previous work about Rootkits in FPGA exists.

19.2 Rootkit Basics

Nowadays, rootkits are a common tool for hackers. They are used to hide attacks from system administrators or antivirus software, and to get further control over the computer. Unlike other forms of malicious code, rootkits hide inside the system to avoid detection and to gain maximal control over the system by becoming a part of the Trusted Computing Base. In the following section we provide an overview about the Trusted Computing Base and rootkits in general.

19.2.1 Trusted Computing Base vs. Trustworthy Computing

The term Trusted Computing Base or TCB is used to describe

> "the set of components (i.e. hardware, software, user) whose correct functioning is sufficient to ensure that the security policy is enforced, or more vividly whose failure could cause a breach of the security policy."
>
> ([4], page 243)

However, having a TCB is not sufficient to provide system security; in addition, the TCB has to be "trustworthy".

For example, if the FPGA is encrypting sensitive data, it is part of the trusted computing base. If the FPGA fails to fulfill this task securely it is still part of the trusted system, but it isn't trustworthy.

The term Trusted Computing has become known to a wider audience through the Trusted Computing Group [5] and their Trusted Platform Module (TPM). The TPM is a chip that implements cryptographic functions like symmetric and asymmetric ciphers, hashing algorithm, random number generators and a secure memory for keys. The TPM is the trusted Platform for security relevant application and as it is implemented in hardware it is harder to attack with respect to software functions. Today the term Trusted Computing is often used as a synonym of a TPM (or a TPM like hardware module).

19.2.2 Modern Rootkits

The most powerful rootkits today are injected into the privileged mode of the processor (Ring 0 in Intel CPU) and, thereby, they have full access to the operating system and the hardware. Common features of today's rootkits include the hiding of processes and files, key logging and hidden data transfers via network. The (almost) total control over the system makes rootkit-detection difficult, even though there is a growing number of antirookit tools available. To avoid detection, some rootkits hide themselves not only in the RAM or on the harddrive but also in the Flash memory of peripheral devices like ACPI-BIOS [6] or PCIcards [7].This makes the detection and the analysis harder. Furthermore, if the system is infected a simple clean-up of the hard disk won't remove the rootkit. A further discussion on that topic can be found in e.g. [8] and [9]. Hoglund also suggested the integration of a rootkit in the CPU microcode, taking rookit development to new level.

19.2.3 Applications of Rootkits in Embedded Systems

While today's rootkits are limited to workstations and servers it is possible that the methods behind them can be use to subvert the Trusted Computing Base of Embedded System

As an example let us consider the following scenario.

Securing the Mediastream in HDTV is a much debated and crucial issue for the success of this technology. Past experience with Pay-TV have shown that securing the content is a hard task. Attackers have found many ways to circumvent the security measures of the TV-stations, forcing them to issue new smart cards or new decoders, eventually leading to a new round of attacks.

Let us assume a company ACME Multimedia Cooperation (AMC) selling multimedia content to end-users via IP.

To protect its valuable content from theft, AMC is using a state of the art crypto protocol, including encryption, authentication and key management. Costumers of AMC can use this content only over a propriety set top box. This box contains among other devices an FPGA responsible for all cryptographic operations.

After some years and many sold boxes, attackers find a vulnerability in the cipher that allows them to decode the multimedia stream.

In a pure hardware implementation of the Cryptosystem it would be almost impossible for the system to recover from this attack. The supplier couldn't implement new cipher algorithm in software because of the required throughput and a change in the Hardware would require a callback of all boxes. As mentioned above, FPGA provides the required computing power for the encryption plus upgrading capabilities to recover from a security breach.

However, a hacker could abuse the FPGA for his own purposes. Such an attack could take place as follows. First the Attacker captures the configuration of the FPGA. In the second step he has to decompile it, then a further analysis is necessary to identify the Achilles heel of the system. The next step is to exploit it by adding or removing functions from the existing configuration. In the last step the modified version of the configuration file is generated and installed at the system.

Eventually the FPGA could transfer the unsecured data stream to a media device, he could bypass the authentication chip card or gather private data about the customer.

The FPGA is still a part of the trusted system at this point, but would not be trustworthy any more.

19.3 Overview of Security Measures in FPGAS

FPGA security research is put on the protection against unauthorized usage of Intellectual Property [10] in the configuration. To prevent this unauthorized usage, several security architectures exist, most of them based on the symmetric encryption of the configuration file in the flash ram and its decryption in the FPGA.

Others prevent the cloning of the configuration file by binding the design to the FPGA using a unique identifier. This identifier can be stored inside a special flash chip with a unique identifier [11,12]. But the flash chip could be replaced by another device. This problem is solved by making the identifier a part of the FPGA, as in case of Spartan 3A with its Device DNA [13]. Packaging the flash memory and the FPGA together (like the Spartan 3AN) is another solution.

A different approach makes use of a non-volatile CPLD to protect the volatile content of the FPGA from cloning [14].

The complexity of the (binary) configuration file and its proprietary (and somewhat secret) format is also seen as a security feature.

19.3.1 Shortcomings of Existing Security Features

Some shortcomings of the existing security architectures are discussed in the following.

19.3.1.1 Data/Configuration Stream Encryption

High end FPGAs like the Virtex series of Xilinx support symmetrical encrypted configurations [17,18]. The encrypted bitstream is stored inside the flash, and decrypted inside the FPGA, making eavesdropping impossible. Cloning is effectively prevented if the key is at least unique to each design. Otherwise an attacker might capture the bitstream of the premium device, and replays it into the low cost version. But there are several problems with the encryption. First of all, only the premium lines (like Xilinx Virtex or Altera Stratix II) offer this option, low cost FPGAs don't have this feature at the moment. Second, as the memory of the FPGA is volatile some designs need a special battery for the key. If the attacker has physical access to the device this might lead to the possibility of a Denial-of-Service attack. Altera overcomes that problem by saving the key in a nonvolatile memory inside the Stratix II [19].

References [18] and [19] describe the used cipher as AES. Since there is no detailed information available beyond this point it is not possible to make further investigation for a wider audience. For example if AES is used in cipher block chaining mode (CBC), the security of the encryption depends strongly on the randomness of the initialization vector (IV). If e.g. a simple counter is used as IV the overall security of the encryption is dramatically reduced.

Further security problems exist for other modes and many times the encryption failed not because of an insecure cipher but an incorrect application of the cipher, as described in [20].

19.3.1.2 Authentication of the Datastream Before Programming

The existing architectures have no strong authentication method for the bitstream, therefore the FPGA cannot determine if the configuration file is valid or not. An attacker can change the configuration stream and thereby change the functionality inside the FPGA. The complex and proprietary structure of the FPGA makes it hard to change the right part of the bitstream. But an attacker might only need to change a couple of bits, with devastating effects to the systems behavior. In [21] the authors described that only 2 bits were sufficient to remove the security control of the Windows NT-Kernel. The attack started with 4 Bytes, as described in [22], and was later reduced to the 2 bits mentioned above. Since FPGA configuration files are at least as complex as i386 machine code it can't be expected that FPGA designs are more robust than the Windows NT machine code.

A solution to this problem is the usage of a strong Message Authentication Code (MAC) for the bit stream. This solution requires an additional secret key to detain an attacker from generating a MAC.

A strong authentication scheme to solve this problem was proposed in [23]. The author recommends the usage of cipher-based MAC (CMAC) with AES as cipher for both encryption and authentication. This approach allows sharing one AES for bit stream encryption and authentication.

19.3.1.3 Authentication of the FPGA Configuration at Runtime

To ensure the integrity of Server-Systems the hashcodes of all important binaries are created after the system is installed or updated. Later on the same procedure is repeated and the new hashcodes are compared to the previously created. A change in the hashcodes indicates that the system has been tampered. At the moment there is no generic solution to check the integrity of the FPGA after the programming. Most FPGAs have a read back functionality that allows a complete dump of the FPGA configuration. However, activating this option offers an attacker the opportunity to obtain sensitive data from the dump. Therefore there is no secure way for the user to verify the integrity of the FPGA because it would invalidate the confidentiality of the system.

A possible solution would be an integrity check that is integrated inside the FPGA and that indicates the system that the configuration is valid or not.

Another solution would be to depend on hardware security only, as described in [4].

19.3.1.4 Configuration/Bitstream Freshness

Replay attacks are very common in computer networks. The attacker is using an old recording of the transmission, for example a successful authentication and retransmits it to the target system. This method could be used to deploy a deprecated and faulty configuration to the FPGA. This security flaw could then be exploited. To carry out this attack, it is necessary to record the communication between the programming device and the FPGA. Later on, this prerecorded communication is transported to the configuration port of the FPGA, bypassing the configuration file inside the flash. This attack works even if the bit stream is encrypted, as the FPGA cannot discriminate between replayed and fresh transmissions. To prevent this kind of attack the FPGA has to ensure that the configuration messages sent are "fresh". This can be done using a nonce (Number once used) that is used for only one transfer. If a replayed attack takes place it will be detected by the FPGA by means of the nonce. The FPGA can then take further steps to protect its integrity e.g. rejection of the bit stream, emergency lockdown, or memory erase.

19.3.1.5 Authorization of the Programming Device

The FPGA input path for programming is a potential security leak itself. Since, the FPGA cannot verify the authorization of the Programming device (Flash, JTAG, or another function), an attacker might replace the original flash ram with a tampered configuration.

FLASH

One possible countermeasure is the usage of flash rams with a unique identifier verified by the FPGA (similar to the solution described in [11]). If the configuration comes from a known and thereby trustworthy flash, it is accepted, otherwise it will be rejected.

JTAG

For the JTAG there is no accepted standard for secure programming and debugging.

Xilinx [24] recommends monitoring the boundary scan communication in the design. In this way the FPGA can detect suspicious commands. The FPGA can either reject/ignore them or can take further steps (e.g. erasing of memory content) to protect the system.

Implementing this security features is totally in the hand of the designer, leaving the risk of incomplete design and faulty implementations.

Microcontroller

When using a microcontroller to program the FPGA the main protection line is the security of the microcontroller itself. If malicious code is deployed into the microcontroller it could, e.g. force a reconfiguration of the FPGA. Since the FPGA cannot discriminate between legal and illegal configuration, it cannot prevent this kind of attack. A white list of valid configurations inside the FPGA would help to minimize this risk by rejecting untrusted configurations.

19.3.1.6 Access Control

Partial Reconfiguration allows the designer to change the configuration [25] of FPGA-components without a complete reprogramming.

There are two flavors of Partial Reconfiguration:

- differential based reconfiguration allows the update of small module parts, and is comparable to a software patch.
- modular reconfiguration allows the swapping of complete modules at runtime.

Some FPGAs allow Designers to control the dynamic reconfiguration by the FPGA itself, using a special on chip component like Xilinxs ICAP (Internal Configuration Access Point) [26,27].

However, existing FPGAs have no hardware solutions to protect important chip areas from reconfiguration. If reconfiguration is enabled the whole configuration can be changed.

Several access control paradigm (as described in [28] and [4]) could be adapted [29] for FPGAs to add defense in depth:

Multilevel Security

For simple applications the LaBella Padula Model (PBPM [4]) could be used to improve the security of the overall system. PBPM works by grouping modules into different security levels. Data exchange takes only place in one direction from lower privileged modules up to modules with higher privileges.

Access Control Matrix

Access Control Matrices store the access privilege of each module to each resource. Therefore, they are only useful for small projects with a limited number of required FPGA resources and HDL modules accessing them. However such an Access Control Matrix could be used to restrict the access to certain, predefined parts of the FPGA fabric. Such a simplified access control could be implemented in hardware and would provide similar security as the protected mode for x86 compatible CPUs.

Role Based Access Control (RBAC)

RBAC [28] is the weapon of choice for most complex software systems today. A RBAC based solution for FPGAs works as follows:

Each HDL-module has one or more roles assigned (e.g. SymetricCipherRole or ProcessorRole). These roles have access privileges (read, write, utilize) to the resources. The advantage of this approach is that roles can be shared among modules, reducing the overhead and enabling the developer to define a clear and stringent security policy from the design process to post mortem analysis of the hardware (by comparing the resource utilization with predefined rules).

19.3.1.7 Security Through Obscurity – Analysis of the Bitstream Format

As mentioned above, the complexity of the bitstream format has been seen as a security feature itself. With modern FPGAs holding more than 1 Million gates, and a programming format kept secret by the vendors, reverse engineering seemed to be a small problem. However this has changed during the last years.

In 2005 the programming format of the ATMEL Atmel AT40k/94k family was posted in the usenet group com.arch.fpga [30]. Based on this work Adam Megacz [31] created a bits a platform for bit stream manipulation and partial reconfiguration.

In 2007 a group of researchers presented a detailed analysis of the configuration format for Xilinx FPGA. They also created a program for analyzing and visualizing compiled bit streams. In [32] they stated that

> "Given a few assumption about the bitstream format … bitstream decompilation becomes both very simple and very fast"

As [33] have shown, cryptographic algorithms can be easily identified in integrated circuits. While to the authors knowledge no such work exist today, for FPGAs future, progresses in this field are highly plausible. Versatile patter recognition tools could analyze the structure of the FPGA, looking for shift registers and other typical sub circuits used. Further information could be gathered from power analysis and other side channel attacks, giving an attacker valuable hints about the location of the wanted circuits. Therefore future synthesize tools might contain netlist obfuscator to hinder such attacks.

19.3.1.8 Power Analysis Attacks Against FPGAs

Power analysis [15] attacks are a based on the observation

> "that the instantaneous power consumption of cryptographic device depends on the data it processes and the operation it performs. [34]"

First Experiments with Power Analysis [16] have shown that FPGAs have no sufficient protection against this kind of attack.

A possible solution to this problem could lie in power analysis aware synthesis tool. These tools might create configurations with constant power consumption, making FPGAs more resistant against power analysis attacks, at the cost of higher resource utilization.

19.4 Rootkits in FPGAS

The security problems mentioned above can be abused to deploy and hide rootkit-like features into an FPGA.

19.4.1 Security Threats Trough Rootkits

The hiding of malicious code (or configuration) inside an FPGA offers an attacker several options to subvert the security of the system.

19.4.1.1 Eavesdropping of Messages

This attack allows listening to confidential messages. Conventional rootkits often have a key logging functionality to listen to passwords, pins and other private information. To obtain such information, a rootkit normally intercepts the communication between the keyboard and the operating system. In FPGA development, wire tapping is a common technique for Designers. Tools like Chipscope pro [35] allow Designers to follow the internal communication of the FPGA. For this purpose a rootkit could use a similar approach to get sensitive system internal information, e.g. unencrypted data streams.

19.4.1.2 Changing of Messages/Leaking of Information

This attack is an extension of the previous one. Here, an attacker does not only listen to the code, but actively changes the messages. This attack is not limited to the content of the message only. The rootkit might change, e.g. timing behavior, power consumption or the electromagnetically emanation of the design to transmit compromising data to the outside world.

19.4.1.3 Replacement/Bypassing of Existing Code-Blocks

If the FPGA is part of the trusted computing base, the outside world relies on the capabilities of the FPGA to enforce its security policy. If an attacker replaces the existing security blocks with, e.g. useless code (either in the memory or in the FPGA itself), it is not able to fulfill the task any more.

19.4.1.4 Denial of service (DoS) / Physical Destruction

As demonstrated in [36] a maliciously crafted bitstream can even destroy a FPGA physically. This could be achieved by:

- connecting an output pin of the FPGA with the output pin of an external device;
- increasing the internal current beyond an acceptable limit, e.g. by combining internal logic blocks.

19.4.2 Deployment of the Rootkit

The three different phases of the life cycle where a rootkit can be deployed are discussed in the following

19.4.2.1 Development Phase

While it seems to be unlikely that a developer would integrate a security related flaw in to the code he writes, it is nonetheless possible, as trusted people don't have to be trustworthy. Other ways a rootkit can be installed are discussed in the following.

Hacking of the Manufacturer's Source-Repository

The intention here is not only to steal the code, but to insert a rootkit into the design. Due to the complexity of Register Transfer Logic Code (e.g. VHDL), the chances are good that this attack would remain unnoticed by the developers. Consequently the rootkit would find its way into the product. Furthermore, auditing of C code for security flaws is more common than, e.g. VHDL or Verilog-audits.

Abusing Hidden Features Inside the Configuration

Sometimes the security flaw is not a bug, but a feature [37]. Vendors could be interested to gather sensitive information too, e.g. for marketing research and integrate hidden functions in the FPGA. Other risks are leftover debug circuits that could be abused by an attacker.

Attacking the Design and Synthesis Tool Itself

A "modified" synthesis tool could insert additional code into the design or alter the netlist, e.g. to allow side-channel attacks. Such kind of attacks would only be noticed by means of a detailed netlist audit. Ken Thompson has stated in "Reflections on trusting Trust" [38]:

> "You can't trust code that you did not totally create yourself. ... No amount of source-level verification or scrutiny will protect you from using untrusted code."

19.4.2.2 Manufacturing Phase

Even modern low cost FPGAs like Xilinx Spartan3E, and S3A, support multiple configuration files. In such systems the rootkit could "sleep" parallel to the normal configuration. An attacker could later force the FPGA to use the "alternative" configuration.

19.4.2.3 "In the Field" Phase

After the system is deployed in the field, an attacker could easily obtain (physical) control over the system. Four different misuse scenarios are presented in the following:

- Change the configuration in the memory followed by a reboot of the system. This could be done by re-flashing the memory or by a physical replacement of the chip (in some cases this "chip" is the flash memory of a compact flash or SD-Card)
- Deploy the rootkit via an onboard JTAG port. This attack would abuse the partial reconfiguration feature of FPGAs described above.
- Abuse the ICAP interface as described above. An attacker could insert his own (assembler) code into the embedded processor using, e.g. a buffer overflow inside a web application [39]. That code could force a partial reconfiguration of the FPGA via ICAP, deploying the rootkit right into the FPGA. E.g. in Embedded Linux [40] the ICAP port is accessible as any other device using the /dev/icap primitive, making an attack at hardware level to a simple cat malicous.bit > /dev/icap after the hacker has reached root privilege.
- A similar attack could be carried out against conventional microcontroller connected with the configuration port of the FPGA.

19.5 Resume and Future Work

Several shortcomings of existing FPGA security architecture where given and possible solutions are proposed. Furthermore we have shown that these security flaws could lead to serious consequences for the overall security of an embedded system.

The injection of a rootkit could happen at any stage of the FPGA lifecycle. Therefore, future work will concentrate on the development of a secure design and development process for FPGAs.

References

1. Instat, FPGA shipments to reach $2.75 bln by 2010, 2006, http://www.instat.com/press.asp?Sku=IN0603187SI&ID=1674. (09.10.2008)
2. T. Wollinger and C. Paar, How secure Are FPGAs in cryptographic applications? (long version),FPL 2004: Proceedings Field Programmable Logic and Applications, 2004, pp. 707–711.
3. J.W. Lockwood et al., An Extensible, System-On-Programmable- Chip Content-Aware Internet Firewall, Field Programmable Logic and Applications (FPL), 2003.
4. R.J. Anderson, Security engineering (Wiley, Indianapolis, 2008).
5. Trusted Computing Group, https://www.trustedcomputinggroup.org/home (accessed 09.10.2008).
6. J. Heasman, "Implementing and Detecting an ACPI Rootkit,", www.blackhat.com/presentations/bh-europe-06/bh-eu-06-Heasman.pdf (09.10.2008).
7. J. Heasman, Implementing and Detecting a PCI Rootkit, www.ngssoftware.com/research/papers/Implementing_And_Detecting_A_PCI_Rootkit.pdf (accessed 09.10.2008).
8. G. Hoglund and J. Butler, Rootkits (Addison-Wesley, Upper Saddle River, 2006).
9. rootkit.com, http://www.rootkit.com/ (accessed 09.10.2008).
10. T. Kean, Cryptographic rights management of FPGA intellectual property cores, FPGA '02: Proceedings of the 2002 ACM/SIGDA tenth international symposium on Field-programmable gate arrays, ACM Press, 2002, pp. 113–118.
11. K. Chapman, Low Cost Design Authentication for Spartan-3E FPGAs, Xilinx Inc..
12. C. Baetoniu and S. Sheth, "XAPP780: FPGA IFF copy protection using Dallas Semiconductor/Maxim DS2432 Secure EEPROM," Xilinx Inc., 2005.
13. K. Chapman, Reading Spartan-3A Device DNA,Xilinx Inc.
14. Altera Corp, FPGA design security solution using MAX II devices, Altera Corp., 2004.
15. S.B. Ors, E. Oswald and B. Preneel, Power-analysis attacks on an FPGA – first experimental results, Cryptographic Hardware and Embedded Systems Workshop 2003.
16. S. Mangard, E. Oswald and T. Popp, Power Analysis Attacks (Springer Science+Business Media, LLC, 2007).
17. C.W. Tseng, Lock Your Designs with the Virtex-4 SecuritySolution, XCell Journal, vol. 52
18. Xilinx,Virtex-5 FPGA Configuration User Guide, Xilinx.
19. Altera Corp., Protecting Intellectual Property through FPGA Design Security, http://www.altera.com/literature/ads/fpgadesignsecurity.pdf (accessed 09.10.2008).
20. H.Wu,The Misuse of RC4 in Microsoft Word and Excel", http://citeseerx.ist.psu.edu/viewdoc/summary?doi=10.1.1.59.5446 (09.10.2008).
21. G. Hoglund and G. McGraw, Exploiting software (Addison-Wesley, Boston, 2004).
22. G. Hoglund, A *REAL* NT Rootkit, patching the NT Kernel, www.phrack.com/issues.html?issue=55&id=5 (accessed 09.10.2008).
23. S. Drimer, Authentication of FPGA Bitstreams: Why and How, Applied Reconfigurable Computing, Springer, 2007, pp. 73–84.
24. G. Crow, Advanced Security Schemes for Spartan-3A/3AN/3A DSP FPGAs, www.xilinx.com/support/documentation/white_papers/wp267.pdf, (09.10.2008).
25. C. Kao, Benefits of Partial Reconfiguration; XCell Journal, vol. 55, 2005.
26. M. Hübner and J. Becker, "Tutorial on Macro Design for Dynamic and Partially Reconfigurable Systems",RC-Education 2006, 2006.
27. Xilinx Inc.Spartan-3 Generation Configuration User Guide, 2006.

28. M. Schumacher, Security patterns (Wiley, Chichester,2006).
29. M. Kucera and M. Vetter, A Generic Framework to Enforce Access Control in FPGAs with Dynamic Reconfiguration, Software Engineering and Applications, ActaPress, 2007.
30. W.S.G. Gosset, "Atmel AT40k/94k Configuration Format Documentation,"2005, http://groups.google.com/group/comp.arch.fpga/msg/a90fca82aafe8e2b (accessed 09.10.2008).
31. A. Megacz, "A library and platform for FPGA bitstream Manipulation," Field-Programmable Custom Computing Machines Symposium, 2007, pp. 45–54.
32. J. Note and E. Rannaud, From the bitstream to the netlist, Departement d`informatique Ecole Normale Superieure, 2007.
33. K. Nohl, D. Evans and H. Plötz, Reverse-Engineering a Cryptographic RFID Tag, USENIX Security Symposium, 2008.
34. S. Mangard, E. Oswald and T. Popp, Power analysis attacks,(Springer, Boston,2007).
35. Xilinx Inc, Chipscope pro, http://www.xilinx.com/ise/optional_prod/cspro.htm (accessed 09.10.2008).
36. I. Hadizc, S. Udani and M.S. Smith, FPGA Viruses, Lecture Notes in Computer Science, vol. 1673, 1999, pp. 291–300.
37. R. Lemos, "World of Warcraft hackers using Sony BMG rootkit," 2005, http://www.securityfocus.com/brief/34 (accessed 09.10.2008).
38. K. Thompson, Reflections on Trusting Trust, Communications of the. ACM, vol. 27, no. 8, 1984, pp. 761–763.
39. A. One, Smashing The Stack For Fun And Profit, Phrack, vol. 7, no. 49, 1996, http://insecure.org/stf/smashstack.html (accessed 09.10.2008).
40. J. Williams and N. Bergmann, Embedded Linux as a platform for dynamically self-reconfiguring systems-on-chip, (09.10.2008), URL: www.linuxdevices.com/articles/AT7708331794.html.

Chapter 20

Bridging the Requirements to Design Traceability Gap

Bernhard Turban[1], Markus Kucera[2], Athanassios Tsakpinis[2] and Christian Wolff[3]

[1]*Electronic Systems Engineering, MBtech Group, Neutraubling, Germany, Bernhard.Turban@micron-ag.com*

[2]*Competence Center SE, University of Applied Sciences, Regensburg, Germany, markus.kucera@informatik.fh-regensburg.de, athanassios.tsakpinis@informatik.fh-regensburg.de*

[3]*Media Computing, University of Regensburg, Germany, christian.wolff@computer.org*

Abstract Requirement traceability ensures that software products meet their requirements and additionally makes the estimation of the consequences of requirement changes possible. In this article a case study analyses symptoms of this problem in the process model of ISO 12207, the foundation of SPICE (ISO 15504), and CMMi. Our analysis is directed at deriving a concept for the integrated extension of current traceability models with the aspect of *documented design decisions*. This integrated decision model is presented along with an additional case study which illustrates the advantages of this approach for traceability.

Keywords Requirements engineering, Traceability, Design, Rationale management, Decision, Embedded systems, SPICE, ISO15504, ISO 12207, CMMi

20.1 Introduction

In the development of safety-critical embedded real-time systems, safety and reliability are of major importance [1] (cf. ISO 61508). Therefore, control and improvement of software processes (cf. ISO 15504 SPICE) are of high significance. In these processes, traceable and consistent elaboration of requirements throughout all development cycles (especially the design phases) is mandatory. However, today's document-heavy approaches face problems with redundancy and synchronization of different stakeholders'

N. Martínez Madrid, R.E.D. Seepold (eds.), *Intelligent Technical Systems*, Lecture Notes in Electrical Engineering 38, DOI 978-1-4020-9823-9_20,

views. To handle these issues, we propose an approach that concentrates on maintaining one consistent view of all requirements between all stakeholders. In the following design phases, the stakeholders and artifacts of the different engineering disciplines (Systems engineering, hardware (HW) and software (SW)) shall be connected by a lightweight model and tool-based traceability approach.

The core of this approach is a decision model which links requirements, design problems and design together. As a result, new constraints on the solution space can be identified and used in a similar way as requirements. Whereas former traceability approaches regarded decisions as valuable side information, in our model decisions get directly integrated in the classical traceability information forming traceable chains of decisions through the design process. As a side effect, the approach addresses several problems in rationale management and encourages direct communication between the stakeholders. This decision model has been integrated into a software development tool which acts as a bridge between requirements tools like DOORS and design-oriented tools like Matlab Simulink or Artisan Realtime Studio.

We start in Section 20.2 with describing the state of the art in traceability research and continue in Section 20.3 to analyze problems in establishing traceability information in current process models. This builds ground for Section 20.4 which introduces our integrated decision model that helps to improve currently used traceability models. A case study shows how the model can be applied in a practical setting. Section 20.5 gives hints on the model's further support potential for designers while Section 20.6 draws a short conclusion.

20.2 Requirement Traceability to Design

Requirements management, i.e. the activity of organizing, administrating and supervising requirements during the whole development process, and *Traceability* are mandatory actions to fulfill exigencies imposed by software engineering standards like SPICE[1] (*Software Process Improvement and Capability determination* [2]) or CMMi (*Capability Maturity Model Integration* [3]). *Traceability* means "comprehensible documentation of requirements, decisions and their interdependencies to all produced information/artifacts from project start to project end" [4 (p. 407)]. Between

[1] In the following, we concentrate on SPICE, but our claims are equally valid for CMMi, as both process models are based on the process model of ISO 12207.

artifacts or respectively models of different development processes emerging structural interruptions and semantic gaps [5,6,7 (p. 138f)] endanger a project's consistency and the common understanding of its stakeholders. Traceability relationships are intended to close these gaps. Paech et al. [8] indicate that traceability in relation to the design of artifacts is typically seen as a set of bidirectional relationships between requirements and their *fulfilling* design entities [9].

Research on traceability has proposed various approaches for establishing or retrieving traceability dependencies. Rochimah et al. present an evaluation of current state of the art traceability approaches concerned with SW evolution [10]. Research has shown that manual creation and maintenance of traceability relations requires enormous effort and includes substantial complexity [11,9,12]. The study of Rochimah et al. further shows that current research on traceability focuses on automating traceability link generation [10(Table 4)]. Some automation approaches still depend on manually established links that are then enriched by supporting automation mechanisms while others are fully automated. We have analyzed the scope of automation of these approaches and identified two major areas of automation:

- Finding interdependencies between different requirements artifacts (e.g. textual documents, use case descriptions, feature-models or analysis models) concerned with requirements.
- Finding interdependencies between design and code artifacts.

Only the approach suggested by Spanoudakis [13,14] tries to establish automated trace links from requirements to models, focusing on analysis models, though.

It is striking that current automated link generation approaches do not concentrate on establishing links between the requirements world and the design world. We believe this can be explained by the "name mapping" or "name referencing" phenomenon: Instead of creating explicit links between items, the same names are used [15 (p. 224)].

If no automatic code generation is available for a design tool and code must be typed manually, traceability must also be established between design and code. As design is (and should be) a more abstract view on the problem modeled, traceability can also be established by naming corresponding elements in design and code identically. This is an explicit heuristic. In addition, another heuristic significantly reinforces this effect in an implicit way: It is very important to achieve a common understanding of the project for all different stakeholders. This can only be achieved, if the project develops a common vocabulary for its used terms. Therefore, in the field of requirements specification, using precise terminology and establishing adequate terminology management is a central principle.

However, concerning traces from requirements to design, Paech et al. [8] point out that these relationships can be of a more complex nature (cf. Fig. 20.1 below). In principle, *non-functional requirements (NFR)* restrain *functional requirements (FR)* and *architectural decisions (AD)*. On the other hand, NFRs are realized by FRs and ADs, whereas FRs are realized and restrained by ADs. Egyed et al. discuss similar observations [11] where they map FRs to nonfunctional aspects (or *software attributes*) where they identify conflicting and supporting situations. It becomes clear that such dependencies are highly dependent on the design context (e.g. the potential conflict can also be nonexistent, if a FR and a nonfunctional aspect are realized in different components).

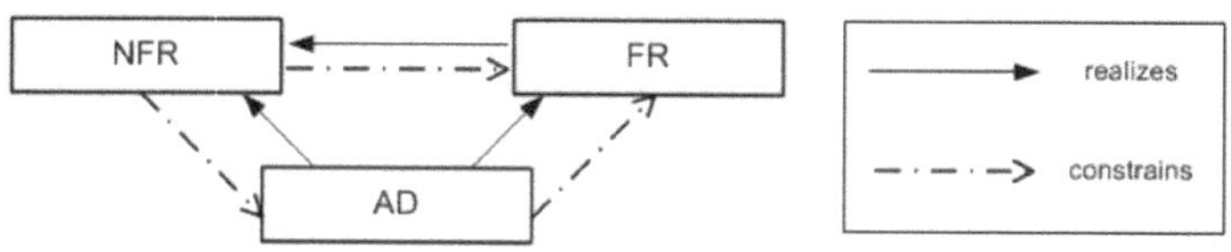

Fig. 20.1 Relationships between non-functional (NFR), functional requirements (FR) and architectural decisions (AD) according to Paech et al. [8]

Tracing requirements from the original requirements specification to design by simple bidirectional links is inaccurate as this would assume the transition from requirements to design to be a fairly linear and one-dimensional process. We rather believe that this transition is a creative and complex mental transfer process performed by designers when gradually transforming the problem space into a solution space (so called *Wicked Problems* [16]). Thus we assume a substantial gap between the world of requirements and design (resp. code), since requirements represent the problem world, whereas design forms the solution world. Accordingly, we believe that (automatic) traceability link generation can be a valuable support mechanism to find dependencies between *within* each sphere (e.g., finding all references of a variable used in source code is a simple and state-of-the-art feature), but it faces high barriers when trying to bridge both worlds. It can be agreed with Egyed et al. that "while some automation exists, capturing traces remains a largely manual process" [17 (p. 115)] and such links degrade over time and must be continuously maintained. Further, the type of usage of the link information must be considered: Egyed et al. [17] distinguish between *short-term utilization* (are all requirements considered?) and *long-term utilization* (assessing a particular change years later). Short-term utilization is more or less covered by the simple link concept usually applied by today's traceability understanding, whereas for mid- and long-term utilization of more complex relations additional information such as decisions and their rationale must be considered.

20.3 Shortcomings of Current Process Artifact Models

The SPICE process model uses the standardized process model of ISO 12207 [2]. This process model demands the following artifacts:

- A system requirement specification (SYS-RS) collects all requirements retrieved from the user by the user requirements specification.
- The SYS-RS builds the basis for a high-level system design model with the prior emphasis on HW-SW-partitioning.
- A HW requirements specification (HW-RS) for the HW and a SW requirements specification (SW-RS) are derived from the SYS-RS and the system design model.
- The HW-RS and SW-RS are the basis for the corresponding HW and SW design models.

We present a detailed analysis of the problems encountered applying traceability to this kind of process model in [18]. In embedded development, requirements concerning the system, SW and HW are strongly interwoven and thus a clear separation between requirements and design artifacts leads to high redundancy and cluttered information. The following example will demonstrate this (a detailed discussion can be found in [18]).

The example has a system requirements specification (SYS-RS) with three requirements causing a problem encountered in our practice context:

- Req.1: An external watchdog component must monitor the system.
- Req.2: Parametric data must be changeable by the customer during operation.
- Req.3: Parametric data must be stored in *Electrically Erasable Programmable Read Only Memory* (EEPROM).

In current practice, the system design determines that the system will include a microcontroller, an external watchdog component and an external EEPROM (cf. Fig. 20.2). The HW requirements specification (HW-RS) derived from the SYS-RS and system design again contains *Req.1* and *Req.3* linking back (fat upward arrows in Fig. 20.2) to the SYS-RS. The detailed HW design determines that watchdog and EEPROM will share the connection pins to the controller by a *Serial Peripheral Interface* (SPI) – communication interface, because other connected components have already used up all remaining pins of the controller. Req.1 gets linked to the watchdog symbol and Req.3 to the EEPROM symbol in the HW design.

The SW requirements specification (SW-RS) contains Req.1, Req.2 and Req.3 linking back to the SYS-RS. During SW design, the architect discovers the potential resource conflict in the shared usage of one SPI for EEPROM and watchdog. Since driving the EEPROM is very time intensive and triggering the watchdog is time critical, the architect rates this combination as a risk, but changes of the HW are rejected due to potentially higher costs. The solution for this conflict, the EEPROM and watchdog drivers must be "artificially" coupled to implement a cooperative handshake solution (Fig. 20.2: association in SW design model marked "!!!"). The solution implies that the planned original standard drivers of a supplier must be adapted internally. In the further progress of the project, these adaptations cause extra efforts not traceable to its background.

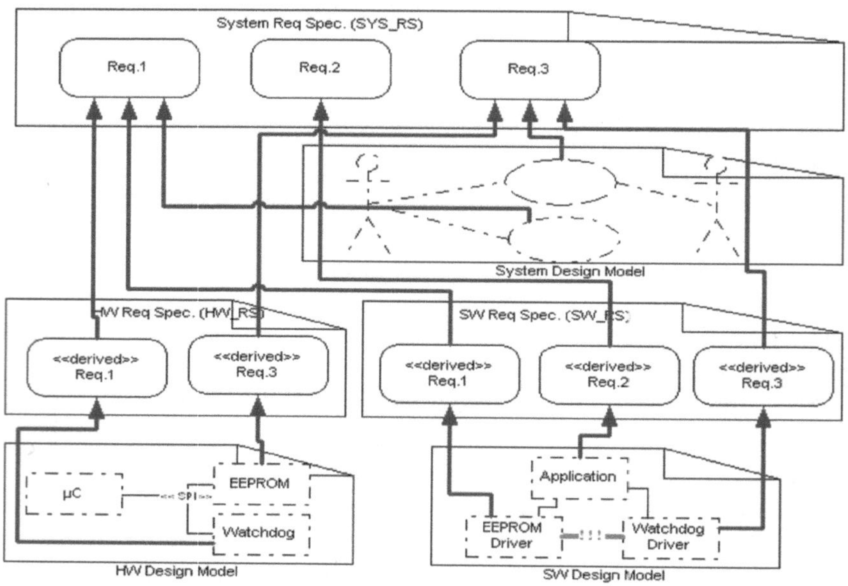

Fig. 20.2 An example following previous approaches

20.4 An Integrated Decision Model

The above example illustrates two central problems: First, the requirements in HW_RS and SW_RS are copies of the requirements in the SYS_RS, leading to high redundancy. In many cases, SW or HW functionality is already clearly demanded for in the user requirements specification. Thus a clear separation of those requirements must be taken over into the SYS_RS and SW_RS respectively HW_RS causing additional effort and redundancies. To avoid this, we propose to use a single central requirements specification containing one consistent view on all aspects of the system to be developed. When a current state of the art requirements management tools like DOORS® is used, a HW-SW-partitioning of requirements is also viable using attributes (proposed values: System, HW, SW, construction, management). Thus, HW-RS as well as SW-RS can be derived as views with a filter on the specific attribute value.

Second, design activities concerning one design artifact (in our example HW design) can have serious implications for other requirement or design artifacts (in our example SW design). This fact is partially considered in the process model of SPICE: System design has high impact on its SW design by raising new "requirements" in addition to the original requirements of the stakeholders. Thus, the idea behind a SW_RS is to collect the SW-related requirements from the SYS_RS and derive new requirements from the System design together. However especially in the automotive sector, SW-design must be subordinated under constraints of extremely cost-optimized HW components. At the moment, SPICE neglects these critical connections between HW and SW.

20.4.1 Introducing the Integrated Model

Another issue in SW requirements which might benefit from more intensive discussion is their negotiability. "Real requirements" are part of the contractual basis between the stakeholders in a project. Changes of such "real" or "contractual" requirements must typically be harmonized with the customer via a *Change Control Board* (CCB) or a similar body used in project management. For requirements resulting from design decisions (modeled as *DesignConstraints* here, see below), it is possible to search for a project internal solution first, before escalating the issue to the CCB is considered. Thus, both kinds of requirements should be strictly separated in their notation.

For this, we propose to use the following taxonomy (Fig. 13.3) to support a more explicit distinction:

- *Requirements* are directly allocated to the SYS-RS, since they concern the legal agreement between customer and contractor.
- "Requirements" derived from requirements or designs are called *DesignConstraints.*
- *Requirements* and *DesignConstraints* have similar qualities and structure. Thus, we use the term *RequirementalItem (RI)* for both items.

Requirements have to refer to their origin [7,4]. This relation should apply to all *RIs*. The origin of *DesignConstraints* lies in previously made design decisions solving the conflicts/forces between *RIs* and/or architectural items constraining the broader more abstract solution space to a more concrete one. These considerations lead to our idea of directly integrating a *decision model* into traceability information (cf. Fig. 20.4) helping to document the origin of new *DesignConstraints* (this especially helps to make the HW' s influence on SW more transparent [19(p. 415)]) in a lightweight and need-oriented way. Figure 20.4 shows this concept extending today's traceability models [8] by an *explicit decision model*. The diagram sketches a concrete situation, where a conflict between two requirements (*Req_1*, *Req_2*) and two UML model elements (*Class1*, *Class2*) is resolved by a *design decision* resulting in two new *DesignConstraints* (*DesConstraint1*, *DesConstraint2*).

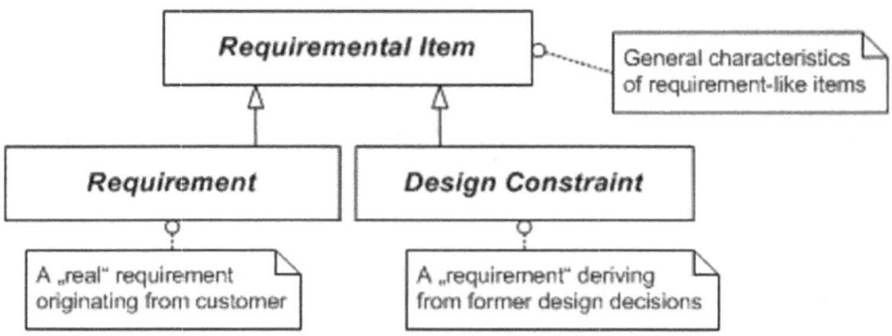

Fig. 20.3 Requiremental items taxonomy

The conventional scheme of relating requirements to realizing model elements is extended by a dialog allowing the capturing of documented decisions. In this dialog, elements of the requirement model and the design model which are conflicting or which cause a problem can be chosen. Equally, diagrams describing aspects of the conflicting situation can be attached as additional information (<<*documenting diagrams*>>).

Furthermore, the decision can be specified on demand via a text component. The text component accepts unstructured text, but may also provide adequate description templates to support the decision documentation. A possible way for structuring this text is shown in Fig. 20.4 with the decision's attributes *assumptions*, *rationales* and *solution specification.*

The decision model presented here is strongly connected to the research area called *rationale management* (RM, cf. [20] for an overview). In [18 (Chapter 5)], we provide a detailed description of the dependencies and implications of research in RM on our decision model.

20.4.2 Applying the Decision Model

The following example illustrates how the same situation as in the example given above is solved by our proposed approach. The system design is done just as proposed in Chapter 3.1 (Fig. 20.5). The SYS-RS contains an attribute that allows a SW-HW partitioning. Req.1 and Req.3 are marked as relevant for HW and SW, Req.2 only for SW. The HW-RS is not directly applied, since the relevant HW requirements are marked in the SYS-RS. The HW design is done similar to Chapter 3.1 and linked to the Req.1 and Req.3 in the SYS-RS. The SW-RS is not applied, since the relevant SW requirements are marked in the SYS-RS. The SW design will be developed from the SYS-RS and the system design model. The architect discovers the same problem concerning watchdog and EEPROM. He opens a decision wizard and marks Req.1 and Req.3 as conflicting and links to the HW-design diagram that documents the conflict. As a further rationale, the architect textually documents "synchronization conflict at SPI between time intensive EEPROM application and time critical watchdog application". A further click helps the architect to put the conflict into the risk list. In the resulting *DesignConstraint*, the architect sketches the cooperative handshake and links the *DesignConstraint* to the EEPROM and watchdog design elements in the SW design.

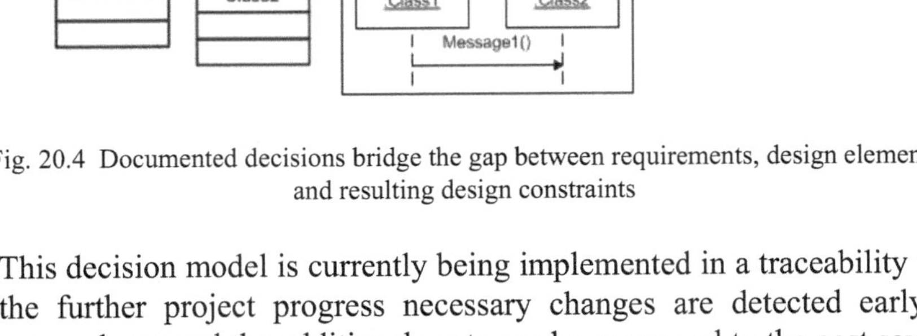

Fig. 20.4 Documented decisions bridge the gap between requirements, design elements and resulting design constraints

This decision model is currently being implemented in a traceability tool. In the further project progress necessary changes are detected early by impact analyses and the additional costs can be compared to the cost savings of the rejected HW change.

The artifacts HW-RS and SW-RS not realized can be generated out of the model, on demand by summing up all requirements related to the corresponding design (HW design model for the HW-RS, SW design model for the SW-RS).

The idea of including decisions into the traceability models is not new (e.g. cf. the recently introduced approach by Tang et al. [21]). In contrast to other approaches that record decisions (rationale) as additional information, our decision model directly integrates into the traceability schema by the following key characteristics:

- Conflicts between *RequirementalItems* (and design elements) can be modeled.
- Decisions do not directly influence dedicated design objects, but they bear *DesignConstraints* that can be the treated as new "requirements" (called *RequirementalItems* here).
- These *RequirementalItems* are part of all subsequent traceability processes.

For a detailed analysis on the differences to other approaches of documenting rationale in design, we recommend reading [18].

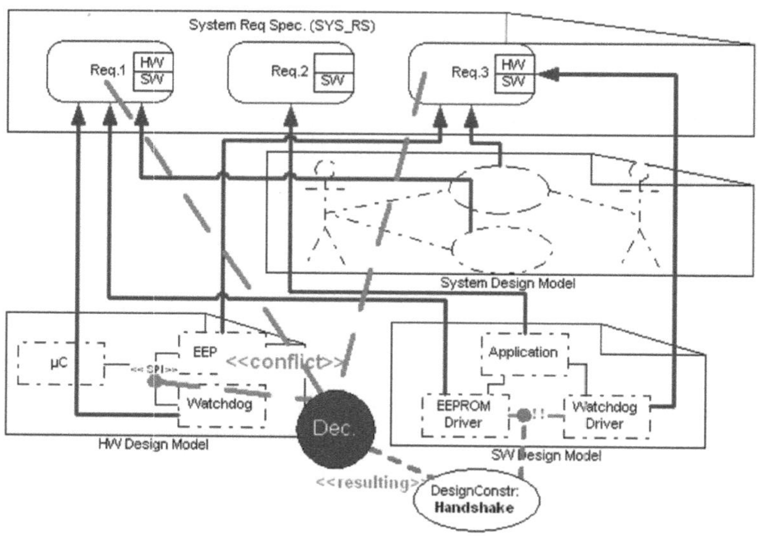

Fig. 20.5 An example following our proposed approach

20.5 How the New Decision Model Provides Additional Support to Designers

In the following, we will discuss additional connections and advantages of the proposed decision model in relation to design-related issues.

20.5.1 Patterns

"Patterns, as used in software engineering, constitute one of the most heavily used approaches for organizing reusable knowledge" [22 (p. 19)]. Patterns define the abstract core of a solution for a continuously recurring problem thus allowing to reapply the solution tailored to the concrete problem [23]. Patterns are described using a structure template. Even though different authors use slightly different templates, the description of the problem (often referred as forces), the solution and its consequences are part of all pattern templates. Our decision model can be described in terms of such a pattern template (see also [24 (Table 1)]): The conflict situation corresponds to the problem description part, whereas the description of consequences in a pattern description could be modeled by resulting new *DesignConstraints*. Due to this analogy, we believe our approach can provide valuable support in selecting design patterns (e.g. the conflict situation of a decision can indicate the usage of a specific pattern). At the same time, it can help knowledge engineers in identifying interesting solutions as *new patterns* (on the relationship between design decisions and patterns also refer to [24,25 (p. 209)]). A pattern library for decisions in modeling embedded systems could be the ultimate goal of such an effort.

18.5 Ensuring Adequate Realization of Design and Decisions

As Posch et al. [25 (p. 38)] underline, architects also have to ensure that their design settings are adequately considered and realized by other designers or coders. Using our model, designers can model the consequences of a decision as *DesignConstraint* and relate the DesignConstraints as new "requirements" (in our terminology: *RequirementalItem* (*RI*)) for design elements. Besides usage in further design or coding processes, the list of assigned *RIs* to a design item can also be used as basis for reviews on design and implementation of the item.

20.5.3 Support for Architecture Evaluation

Our approach can also provide valuable support at maintenance and evaluating architectures [26]. As Moro [27 (p. 321)] points out the usage of patterns and other decisions must be documented for later maintenance and architecture evaluation issues.

20.6 Summary and Outlook

This article shows the interdependencies between the SPICE-layered process model, requirements, traceability, designs and decisions with special attention on low redundancy in the traceability information. We suggest a strict separation between contractual mandatory requirements ("real requirements") and requirements resulting from former design decisions (*design constraints*). Design decisions are interpreted as links between requirements, designs and derived *DesignConstraints*. This closely connects and synchronizes approaches in requirement traceability and rationale management. In accordance with the literature [28,6,8,29,12], it can be argued that the influence of requirements on design processes – and vice versa – is only insufficiently modeled by bidirectional linkages.

In the course of a cooperation project between *MBtech Group* (formerly with the *Micron Electronic Devices AG*, since June 2008 part of *MBtech*), the *Competence Center for Software Engineering* of the *University of Applied Sciences Regensburg* and *the Media Computing Group* of the *University of Regensburg* a prototype system is being implemented which includes the decision model presented here. Customer workshops at *MBtech* have shown promising acceptance by designers. At the moment, the tool environment faces first practical applications in real world projects.

Acknowledgements This research has been funded by the Bavarian Ministry of Economic Development (Grant Nr. IUK229). Furthermore, we want to thank all partners that contributed to our research.

References

1. O. Benediktsson, R. Hunter and A.D. McGettrick. Processes for Software in Safety Critical Systems. In: Software Process: Improvement and Practice 6 (1), 47–62 (2001).
2. K. Hörmann, L. Dittmann, B. Hindel and M. Müller. SPICE in der Praxis, Interpretationshilfe für Anwender und Assessoren, dpunkt Verlag, Heidelberg (2006).
3. R. Kneuper. CMMI. Verbesserung von Softwareprozessen mit Capability Maturity Model Integration. Volume 2, dpunkt Verlag, Heidelberg (2006).
4. Ch. Rupp. Requirements-Engineering und –Management, Volume 2, Hanser, München (2002).
5. M. Lindvall. A study of traceability in object-oriented systems development. Licenciate thesis, Linköping University, Institute of Technology, Sweden (1994).
6. A. von Knethen. Change-Oriented Requirements Traceability. Support for Evolution of Embedded Systems, Fraunhofer IRB Verlag, Stuttgart (2001).
7. Ch. Ebert. Systematisches Requirements Management, dpunkt, Heidelberg (2005).

8. B. Paech, A. Dutoit, D. Kerkow and A. von Knethen. Functional requirements, non-functional requirements, and architecture should not be separated – A position paper, REFSQ Essen (2002).
9. O. Gotel, O. and A. Finkelstein. An Analysis of the Requirements Traceability Problem. Proceedings First International Conference on Requirements Engineering 1994, pp. 94–101 (1994).
10. S. Rochimah, W. Wan Kadir and A. Abdullah. An Evaluation of Traceability Approaches to Support Software Evolution. International Conference on Software Engineering Advances (ICSEA) (2007).
11. A. Egyed and P. Grünbacher. Indentifying Requirements Conflicts and Cooperation: How Quality Attributes and Automated Traceability Can Help. IEEE SW November/December (2004).
12. B. Ramesh and M. Jarke. Toward Reference Models for Requirements Traceability. IEEE Transactions on Software Engineering, 27(1) (2001).
13. G. Spanoudakis, A. Zisman, E. Perez-Minana and P. Krause. Rule-Based Generation of Requirements Traceability Relations, Journal of Systems and Software, 105–227 (2004).
14. G. Spanoudakis, "Plausible and Adaptive Requirement Traceability Structures," in Proc. 14th International Conf. Software Eng. and Knowledge Eng. (2002).
15. M. Müller, K. Hörmann, L. Dittmann and J. Zimmer. Automotive SPICE in der Praxis: Interpretationshilfe für Anwender und Assessoren. Dpunkt 1. Auflage, Heidelberg (2007).
16. W. Kunz and H. Rittel. Issues as elements of information systems. Working Paper 131, Center for Urban and Regional Development, University of California, Berkeley (1970).
17. A. Egyed, P. Grünbacher, M. Heindl and S. Biffl, Value-Based Requirements Traceability: Lessons Learned. 15th IEEE International Requirements Engineering Conference (2007).
18. B. Turban, M. Kucera, A. Tsakpinis and Ch. Wolff, An Integrated Decision Model For Efficient Requirement Traceability in SPICE Compliant Development, Fifth Workshop on Intelligent Solutions in Embedded Systems (WISES), Madrid (2007).
19. P. Liggesmeyer and D. Rombach (Eds.): Software Engineering eingebetteter Systeme Grundlagen – Methodik – Anwendungen. Volume 1., Elsevier, München (2005).
20. A. Dutoit, A., R. McCall, I. Mistrik and B. Paech (Eds.). Rationale Management in Software Engineering. Springer, Berlin (2006).
21. A. Tang, Y. Jin and J. Han, A rationale-based architecture model for design traceability and reasoning. Journals of Systems and Software Volume 80(6) 918–934. (2007).
22. A. Dutoit, R. McCall, I. Mistrik and B. Paech. Rationale Management in Software Engineering: Concepts and Techniques. In [20 (p.1–48)] (2006).
23. E. Gamma, R. Helm, R. Johnson and J. Vlissides, Design Patterns: Elements of Reusable Object-Oriented Software. Addison-Wesley, Reading, MA (1995).
24. N.B. Harrison, P. Avgerion and U. Zdun, Using Patterns to Capture Architectural Decisions. IEEE Software 38–45 July/August (2007).
25. T. Posch, K. Birken and M. Gerdom, Basiswissen Softwarearchitektur- Verstehen, entwerfen, bewerten und dokumentieren. dpunkt, Heidelberg (2004).
26. P. Clements, R. Kazman and M. Klein, Evaluating Software Architectures – Methods and case studies. Addison-Wesley, New York (2002).
27. M. Moro, Modellbasierte Qualitätsbewertung von Softwaresystemen, Books on Demand GmbH, 1. Auflage (2004).

28. A. von Knethen, A Trace Model for System Requirements Changes on Embedded Systems, In Proc. of 4th International Workshop on Principles of SW Evolution, Sept. (2001).
29. R. Pettit, Lessons Learned Applying UML in Embedded Software Systems Design, Second IEEE Workshop on Software Technologies for Future Embedded and Ubiquitous Systems, Wien (2004).

Kolibri 360 GmbH
Ferdinand-Jühlke-Straße 7
99095 Erfurt, Deutschland
produktsicherheit@kolibri360.de